science BASICS

Joanne Macown

04 Year 10

ACIDS & BASES

ELECTRICITY

Earth's Changing Face

GENES & BABIES

Australia • Brazil • Japan • Korea • Mexico • Singapore • Spain • United Kingdom • United States

Science Basics. 4, Year 10
1st Edition
Joanne Macown

Text designer: Fran Whild
Text illustrator: Richard Gunther
Cover designer: Brenda Cantell
Production controller: Carlie Devine
Typeset in 10/14 Rotis Sans Serif by BookNZ

National Library of New Zealand Cataloguing-in-Publication Data
Macown, Joanne.
Science basics. 4, Year 10 / Joanne Macown.
ISBN 978-0-17-013130-8
[1. Science—Problems, exercises, etc.—Juvenile literature.]
1. Science—Problems, exercises, etc. I. Title.
507.6 —dc 22

Cengage Learning Australia
Level 7, 80 Dorcas Street
South Melbourne, Victoria Australia 3205

Cengage Learning New Zealand
Unit 4B Rosedale Office Park
331 Rosedale Road, Albany, North Shore 0632, NZ

For learning solutions, visit **cengage.com.au**

Printed in Australia by Ligare Pty Limited.
9 10 11 12 13 14 15 15 14 13 12 11

Contents

UNIT 01 HUMAN VARIATION

All of the members of your class are about the same age. You are all members of the human species. Yet you all have different characteristics. This makes you an individual. Your unique features make you a person other members of the class recognise as being you.

The differences between individuals are called variations. Some of these variations come from the genetic information you get from your parents. They are called inherited variations. There are two main types of inherited variation.

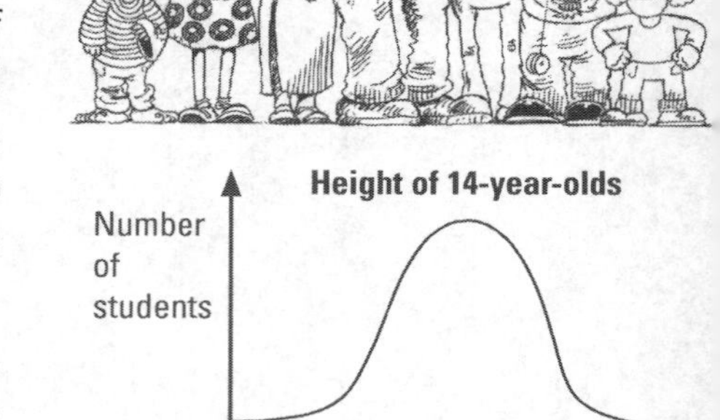

1 Continuous variation: These characteristics show a wide range of possibilities that blend into one another, such as height, colour of hair and eye colour of 14-year-olds. When a continuous characteristic is measured in a large group of people, then plotted on a frequency graph, the graph line has a 'bell shape'. This bell shape is called a normal distribution curve. The same shape appears whenever the frequency of a continuous variation is plotted.

Height of 14-year-olds

Number of students

1.40 m Height 1.8 m

2 Discrete (either/or) variation: Either you have these characteristics or you do not. There is no range of possibilities. There is no blending of one characteristic into the next. Examples of human 'either/or' characteristics are:

Either	Or	Either	Or
tongue roller	non-roller	free ear lobe	fixed ear lobe
hitch-hiker's thumb	straight thumb	widow's peak	straight hairline
cleft chin	straight chin	right-handed	left-handed

Variations are also a result of the environment affecting your inherited characteristics. These are called acquired variations. For example, a child who inherits the light skin characteristic from his or her parents will have light skin. If that child grows up in a country with a hot, sunny climate, the skin will darken. This is because in sunlight the skin produces a brown pigment called melanin. Melanin provides some protection from sunburn. When the child grows up and has children, these children will not be born with dark skin. This is because dark skin has been acquired, not inherited.

People are the result of interaction between what they inherit (called nature) and their environment (called nurture). Nurture is not just the physical environment but how a person is raised. How well you do at school is not just a result of intelligence characteristics you inherited from your parents (nature). How well you do at school is affected by how your family and your teachers encourage you and how well you work at succeeding (nurture).

Variation occurs in all living things, not just humans. Think about the variety of dogs, cats, apples, potatoes and roses there are. Having a range of different characteristics is important if a species is going to survive. In most species of plants and animals the individuals with the 'best' characteristics for a particular environment reproduce and pass their characteristics on to the next generation. When the environment suddenly changes, some individuals will be better suited to the new environment because of their differences and they will survive. Sometimes different characteristics help an individual out-compete others for things like food, water and shelter. Variation has an effect on evolution (how a species changes over time).

4U2DO

1 Describe two features you have inherited from

a) your mother. ______

b) your father. ______

2 Describe any features that have not been inherited from your parents (for example tattoos or scars).

______

3 Your characteristics

a) On the outline of the face, draw in the 'either/or' features you have (from those listed in the table on page 2).
b) Circle which features in each pair you have – hitch-hiker's thumb/straight thumb, right-handed/left-handed.

4 Continuous variation

a) Name two features you have that are examples of continuous variation.

______

b) Why is it hard to accurately describe the colour of your hair?

______

5 Name

a) three breeds of dog. ______

b) three varieties of apples. ______

6 Why is it important to have variety in species of living things?

______

7 Write down three things you have learned from this unit.

a) ______

b) ______

c) ______

8 Write down anything you need to ask your teacher to explain. ______

UNIT 02 Chromosomes, Genes and DNA

(1) Your body is made of cells. The nucleus is the control centre of each cell.

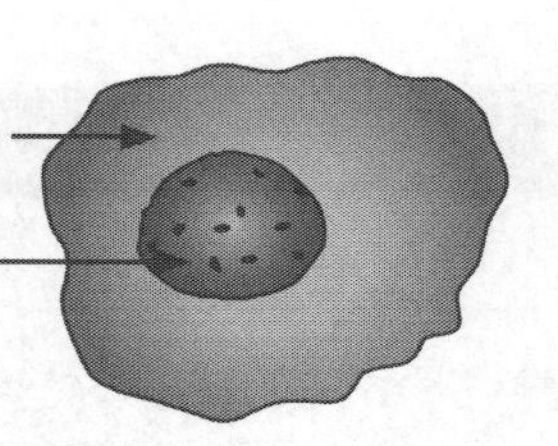

(3) Chromosomes are made up of genes. There are hundreds of genes on each chromosome. Genes are strings of chemicals that help create the proteins that make up your body. Proteins determine the colour of your eyes, the shape of your ears and whether you will be tall or short as well as how your cells and body work.

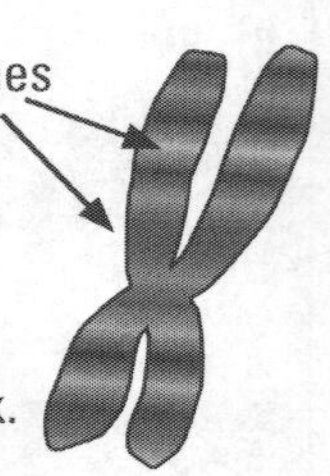

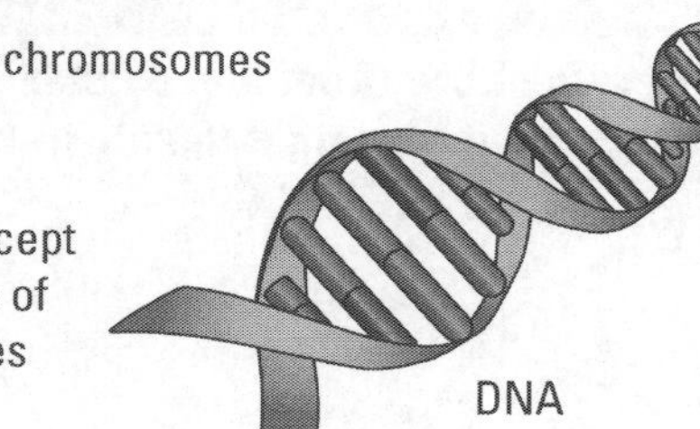

(2) Every nucleus in your cells (except the sex cells) contains 23 pairs of chromosomes (46 chromosomes in total).

(4) Genes are made up of a chemical called deoxyribonucleic acid (DNA). DNA is a large chemical molecule in the shape of a twisted ladder or double helix. The rungs of the ladder are made up of a pair of chemical 'bases'. There are only four bases in the DNA 'alphabet'. Each 'DNA word' is three bases long. A gene is a genetic 'sentence' made up of many of these three-lettered words.

(5) Your chromosomes came from your biological parents.

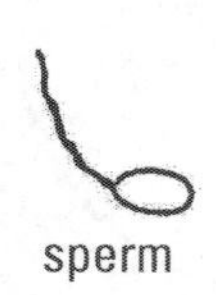

egg

(6) A special type of cell division produces sex cells (eggs and sperm). These sex cells contain only 23 chromosomes. One member of each pair found in a normal body cell is chosen at random to go into the sex cell so they carry a complete set of information.

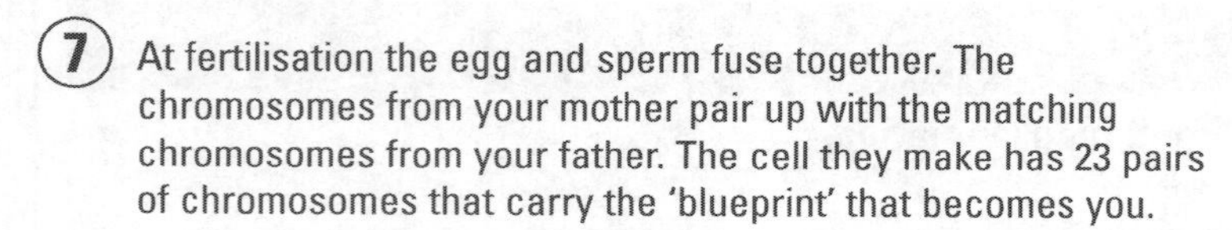

(7) At fertilisation the egg and sperm fuse together. The chromosomes from your mother pair up with the matching chromosomes from your father. The cell they make has 23 pairs of chromosomes that carry the 'blueprint' that becomes you.

Because you have a mixture of chromosomes from both your mother and your father, you do not look exactly the same as either one of them. You will have some characteristics from each of them, for example your mother's eyes and your father's nose, but you are an individual. You will have some characteristics in common with your brothers and sisters because you have the same parents, but unless you are an identical twin, you will not be exactly the same as them.

Boy or girl?

One noticeable characteristic of humans is that some of us are female and some of us are male. This characteristic is inherited from our parents. There is one pair of chromosomes (pair number 23) that determines if you are a boy or a girl.

The father is the parent that determines whether a child is male or female. This is because it is the father who passes on the Y chromosome.

Women have two X chromosomes at pair 23. The eggs they produce all carry one X chromosome. So they always pass an X chromosome on to their children.

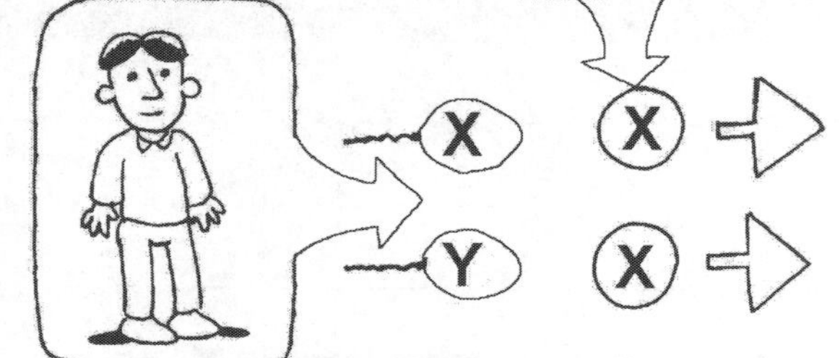

Men have one X chromosome and one Y chromosome at pair 23. Half the sperm they produce carry an X chromosome and half carry a Y chromosome.

If a sperm carrying an X chromosome fertilises an egg, the child that develops will be a girl (XX chromosomes).

If a sperm carrying a Y chromosome fertilises an egg, the child that develops will be a boy (XY chromosomes).

1 Label the diagram using these words: gene, nucleus, chromosome, DNA, bases.

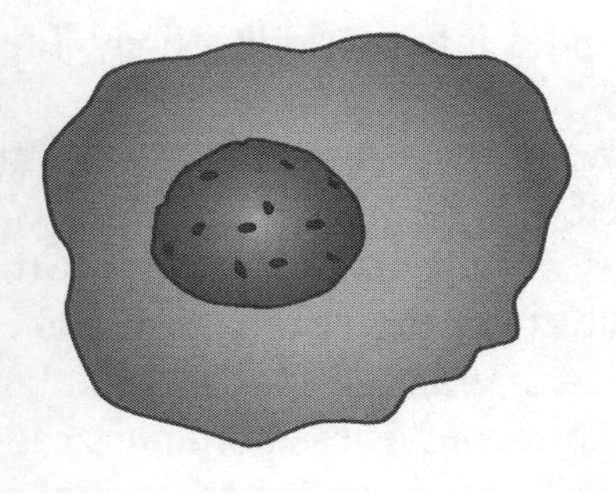

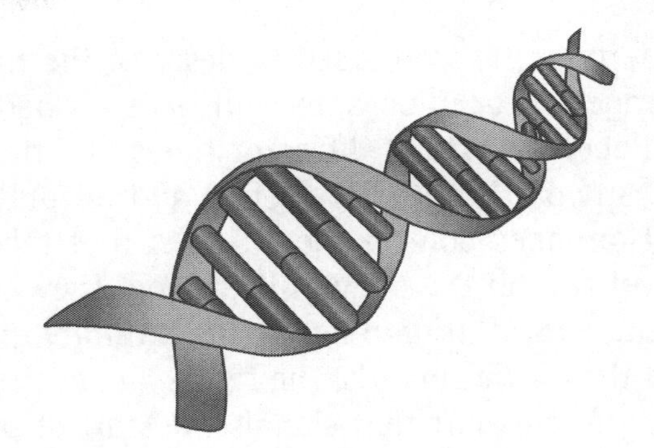

2 Imagine all the chromosomes in the nucleus of a cell were the 'Owner's Manual' for your body telling it how to work and what features to have. Fill in the missing words to complete the comparison.

a) Every chromosome is a ______________________ in the manual.

b) Several genes make up a ______________________ in a chapter.

c) Each individual gene is a ______________________ in the paragraph.

d) Every three bases makes up a ______________________ in the sentence.

MISSING WORDS	
word	sentence
paragraph	chapter

3 Why do you have some but not all features in common with your brothers and sisters?

____

____

4 Answer the following.

a) Why do all boys contain an X chromosome?

____

b) Why is it the father who determines the sex of any baby?

____

c) Every time a woman gets pregnant, what is the percentage chance she will have a girl?

____

5 Write down three things you have learned from this unit.

a) ____

b) ____

c) ____

6 Write down anything you need to ask your teacher to explain.

____

____

UNIT 03 PUBERTY

Puberty is the word used to describe the time during which a child's body changes into an adult's body. These changes occur in boys and girls so their bodies are ready to reproduce.

Puberty starts at different times in different people. It generally starts between the ages of 10 and 15 and takes two or three years before all the adult characteristics have fully developed.

Hormones control the changes that take place during this time. Hormones are chemical messengers that travel around the body in the blood. They are made in one part of the body but affect target organs elsewhere in the body. During puberty, the pituitary gland in the brain sends out hormones that act on the testes in boys and the ovaries in girls. The testes and ovaries then begin to produce their own hormones. These hormones cause the growth spurt that signals the start of puberty. The hormones also cause the development of secondary sex characteristics. The diagrams show what these secondary characteristics are in girls and boys.

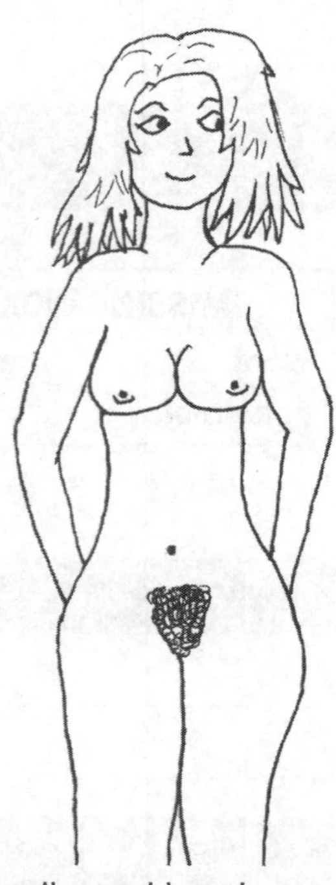

- body grows taller and heavier
- sex hormones (oestrogen and progesterone) are made so sex organs mature
- sex organs get bigger ready for sexual intercourse
- breasts get larger ready to produce milk
- hips widen so a baby can fit through the pelvis during birth
- eggs (or ova) are produced by ovaries ready to be fertilised by a sperm
- menstruation (periods) start
- pubic hair grows around external sex organs (genitals)
- body hair grows under arms

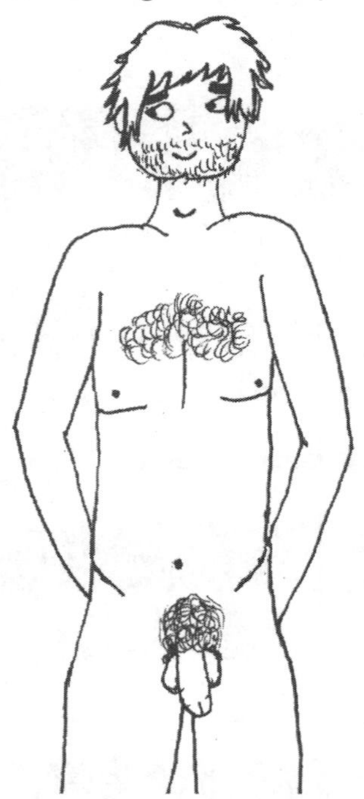

- body grows taller and heavier
- sex hormones (testosterone) are made so the sex organs mature
- sex organs get bigger ready for sexual intercourse
- sperm cells are produced ready to fertilise an egg
- voice box (larynx) gets larger and voice deepens
- body hair grows under arms and on the chest
- facial hair starts to grow
- pubic hair grows around external sex organs (genitals)

Many people suffer from pimples during puberty. This is because of the hormone changes that occur in the body during this time. One hormone causes oil glands in the skin to produce oil, often too much oil. The oil blocks the pores on the surface of the skin causing a blackhead or whitehead.

Bacteria living and reproducing in the oil causes the swelling, redness and pus you get with a pimple. Pimples should not be squeezed because that can cause scarring. Regular washing and the use of some medicated face-washes help prevent pimples, but they cannot cure pimples because the basic cause is hormones. Once the hormone balance is reached, pimples generally stop occurring.

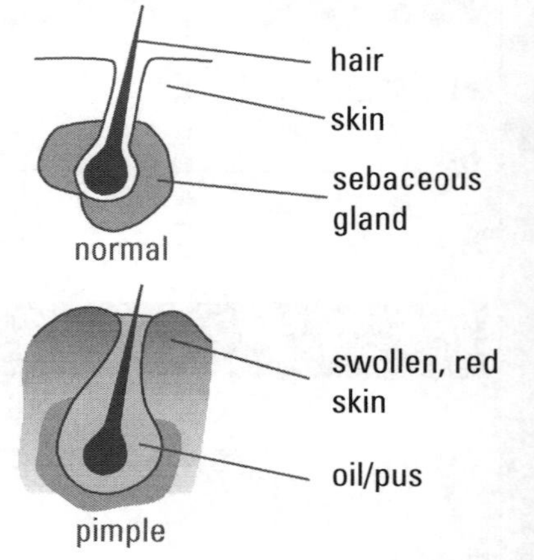

1 How does the brain control the changes that take place at puberty?

2 Hormones

a) What are hormones?

b) How do hormones get around the body?

3 Why are the changes that happen during puberty called the secondary sex characteristics?

4 Complete the table to show the changes that happen to girls only, to both boys and girls, and to boys only.

Girls only	Girls and boys	Boys only

5 Suggest some reasons why puberty starts at different times for different people.

6 Puberty makes the body physically mature and ready for reproduction. How else do people need to mature before they are ready to be parents?

7 Write down three things you have learned from this unit.

a)

b)

c)

8 Write down anything you need to ask your teacher to explain.

UNIT 04 THE MALE REPRODUCTIVE SYSTEM

Humans reproduce sexually. This means a sex cell from a man (a sperm) must combine with a sex cell from a woman (an egg or ova) if a baby is going to develop. The male reproductive system is designed to

- produce male sex cells called spermatozoa (sperm for short)
- produce the male sex hormone called testosterone
- keep these sperm alive until they can be placed into a woman's body
- get the sperm into the woman's body.

Sperm are produced continuously throughout a man's life starting from puberty. A normal, healthy man produces over 100 million sperm each day. They are stored until released from the man's body during either sexual intercourse, masturbation or a wet dream. The release of sperm from the penis is called ejaculation. Normally there are between 200 and 600 million sperm released in each ejaculation.

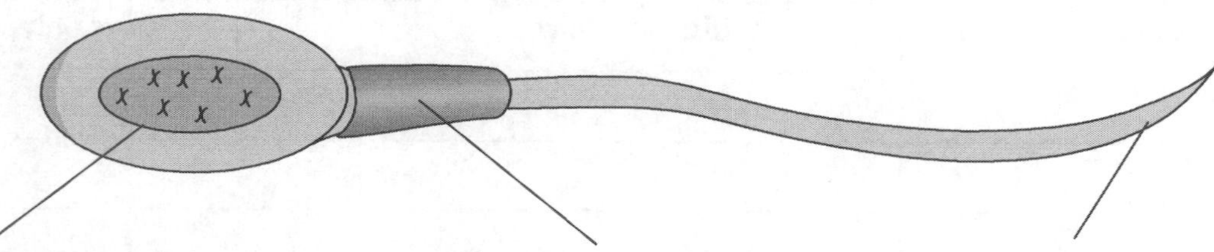

- The 'head' of a sperm cell has a nucleus containing 23 chromosomes from the father. It also has a structure at the tip to help the sperm enter the egg.
- The midpiece contains cell structures that make the energy the tail needs to move.
- The 'tail' is used to swim towards the egg once the sperm is inside a woman's vagina.

Most of the male reproductive system is outside the body and is easy to see. The diagram shows inside both the external genitals and internal sex organs and explains what their job is.

Side view of external and internal male sex organs

- urethra – a tube that carries semen through the penis before being released from the body. The urethra also carries urine from the bladder out of the body but it can only do this when the penis is not erect.
- penis – a tube-shaped organ made of spongy tissue. Blood fills the spongy tissue when a man is sexually aroused, making the penis stiff and erect so it can be placed into the woman's vagina during sexual intercourse.
- prostate gland and seminal vesicle – produce fluids that help the sperm swim and 'feed' the sperm so they survive. These fluids, mixed with the sperm, produce a white fluid called semen.
- sperm tube – carries the sperm from the testes to the penis.
- epididymis – a coiled tube where sperm are stored while they mature.
- testes (or testicles) – two oval balls that produce sperm and the male hormone testosterone. They are found in a bag of skin called the scrotum, which is outside of the abdomen so is cooler by 2–3°C than normal body temperature. This cool temperature is necessary for sperm production.

When baby boys are born, the end of the penis (the glans) is covered by a fold of skin called the foreskin. For cultural, medical or religious reasons, this foreskin is surgically removed from some boys when they are only a few days old. This operation is called circumcision. Some research suggests circumcised males are less likely to suffer from infections because the glans is easier to clean.

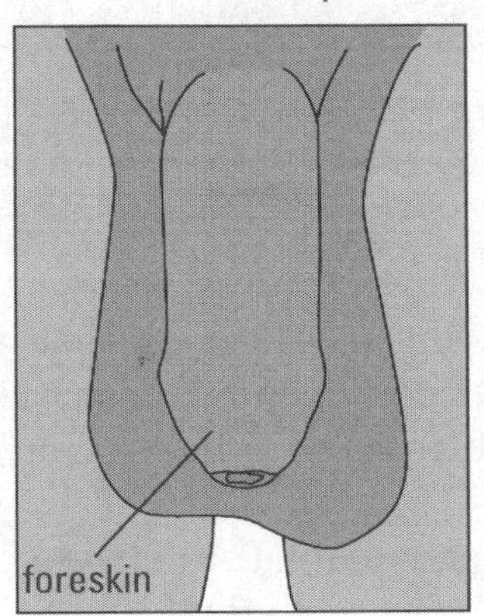

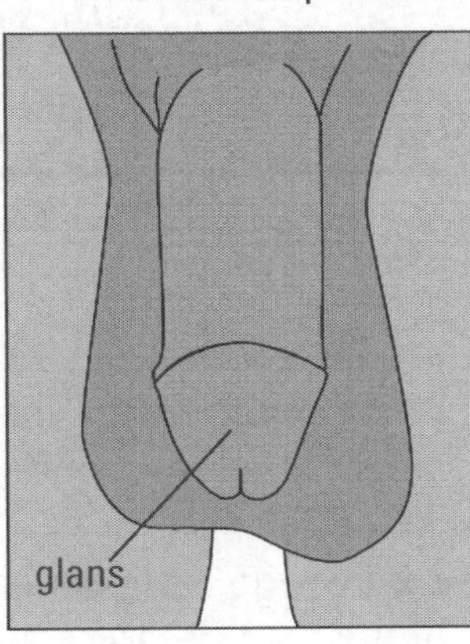

4U2DO

1 What are two purposes of the male reproductive system?

2 What causes an erection?

3 Describe the journey sperm take when they are released from a man's body.

4 What is semen and what is its job?

5 Why are the testes in the scrotum hanging outside of the abdomen?

6 What are the advantages of having two testes instead of just one?

7 Write down three things you have learned from this unit.

a)

b)

c)

8 Write down anything you need to ask your teacher to explain.

UNIT 05 The Female Reproductive System

The female reproductive system is designed to

- produce female sex cells called eggs or ova (singular is ovum, one egg)
- produce the female hormones oestrogen and progesterone
- carry the egg to where it can combine with a sperm (this joining of egg and sperm is called fertilisation)
- receive the male penis and sperm during sexual intercourse
- nourish and protect a baby before and after birth.

Girls are born with all the eggs they will ever produce already in their ovaries. A fluid-filled envelope of cells called the follicle surrounds each egg. These protect and nourish the egg. There are about 400,000 follicles per ovary but only several hundred will release their eggs during a woman's reproductive years. At puberty, hormones cause one follicle each month to mature and release its egg into the oviduct.

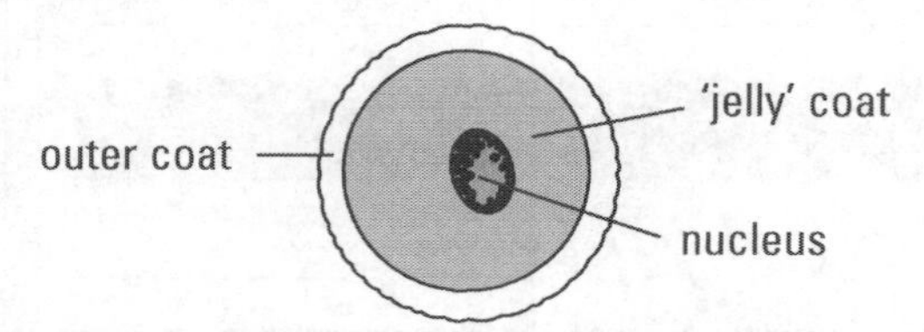

In the nucleus of each egg are 23 chromosomes from the mother. A single egg is about the size of the full stop at the end of this sentence. It is about 2000 times larger than a sperm cell and is larger than any other cell in a woman's body.

The external sex organs (genitals) in the female are called the vulva. They are harder to see than the male genitals because they are between a woman's legs.

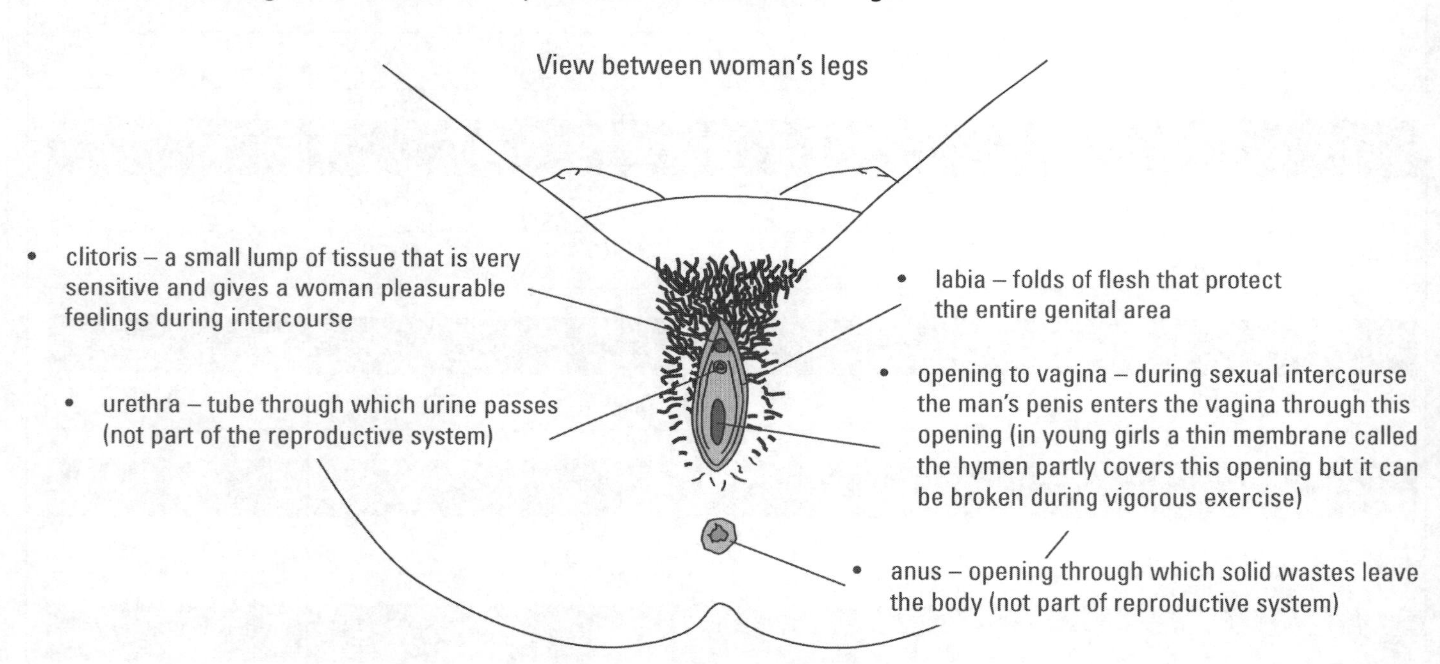

Most of the female reproductive system is inside the body. The diagram on page 11 shows the inside of the internal sex organs (as seen from the front) and explains what their job is.

The breasts or mammary glands are designed to produce milk to feed a baby once it has been born. They are made of fatty tissue and milk-producing glands. The nipple has tiny openings for the milk to come out of, and a darker area called the areola surrounds it.

Unlike men, who produce sperm from puberty until they die, women only produce eggs from puberty until between the ages of 45 and 55. Menopause is the term used to describe the time in a woman's life when she stops producing eggs and stops getting her period (see Unit 6).

- oviduct (or fallopian tube) – a very narrow tube about 10 cm long with an opening that looks like a frilly-edged funnel, carries the egg to the uterus (or womb). It is where the egg and sperm usually meet.

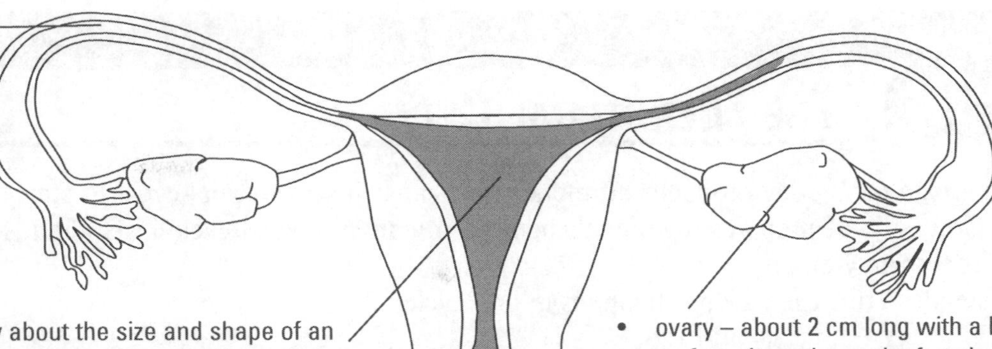

- uterus (or womb) – normally about the size and shape of an upside-down pear, it has a muscular wall and an inner lining with a rich blood supply. This blood supply nourishes the baby as it grows during a woman's pregnancy.
- ovary – about 2 cm long with a bumpy surface, it produces the female sex cells (eggs) and the female hormones oestrogen and progesterone.
- vagina – a tube 8–10 cm long with thin, muscular walls that have glands that produce a fluid to keep it clean and moist. During intercourse the man's penis enters the vagina and sperm are released near the cervix. During birth, the baby passes through the vagina as it leaves the mother's body.
- cervix – a ring of muscle at the bottom of the uterus that opens into the vagina.

4U2DO

1 What are three purposes of the female reproductive system?

____

2 Why do you think most of a woman's reproductive organs are inside her body?

____

3 What does the word vulva refer to?

____

4 What are the advantages of having two ovaries instead of just one?

____

5 Why do the uterus and the vagina have muscular walls?

____

6 Why do you think a woman produces only one egg at a time but men produce millions of sperm in each ejaculation?

____

7 Write down three things you have learned from this unit.

a) ______________________________

b) ______________________________

c) ______________________________

8 Write down anything you need to ask your teacher to explain.

____

____

UNIT 06 The Menstrual Cycle

At puberty, a girl's body produces hormones that cause her menstrual cycle to start. The menstrual cycle is the process that prepares a woman's body for pregnancy. Each cycle lasts about 28 days but it varies from woman to woman.

In simple terms this is what happens in one cycle:

- An egg matures in the ovary and is released into the oviduct.
- The egg moves towards the uterus. It may meet a sperm in the oviduct along the way.
- The lining of the uterus thickens in preparation for the arrival of the embryo that will form if the egg is fertilised by a sperm.
- If the egg is not fertilised, the uterus lining breaks down and passes out through the vagina carrying the egg with it. When this bleeding occurs, we say a woman is menstruating or has her 'period'.

pituitary gland

FSH

The table shows in more detail what happens during one average cycle of 28 days. The first day of a woman's 'period' is Day 1 of her cycle.

Days of the cycle	Hormones involved at each stage of the cycle	What happens at each stage of the cycle	
1–5	Follicle stimulating hormone (FSH) from the pituitary gland in brain affects the ovary.	An egg starts to mature in a follicle in one of the ovaries.	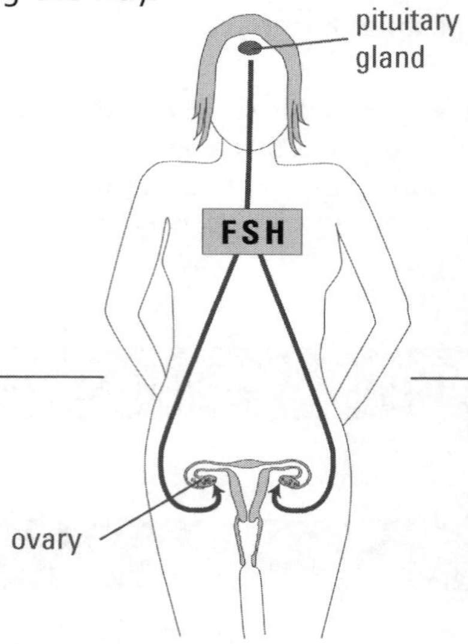 ovary
6–12	Oestrogen is produced by the ovary and has an effect on the uterus.	A new uterus lining starts to grow.	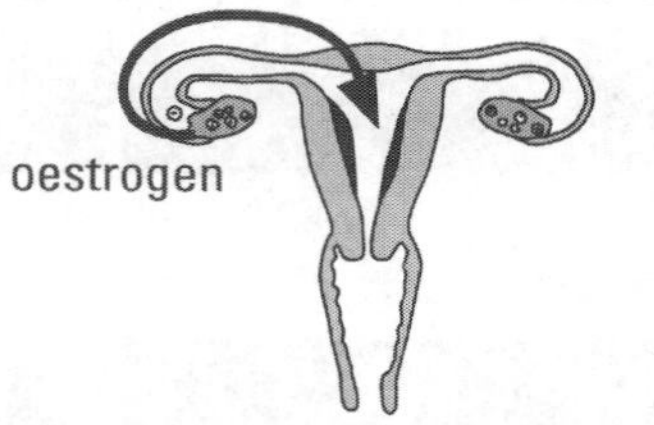 oestrogen
13–15	Oestrogen reaching the pituitary stops it making FSH and starts it making lutenising hormone (LH).	The follicle bursts, releasing the egg into the oviduct. This is called **ovulation**.	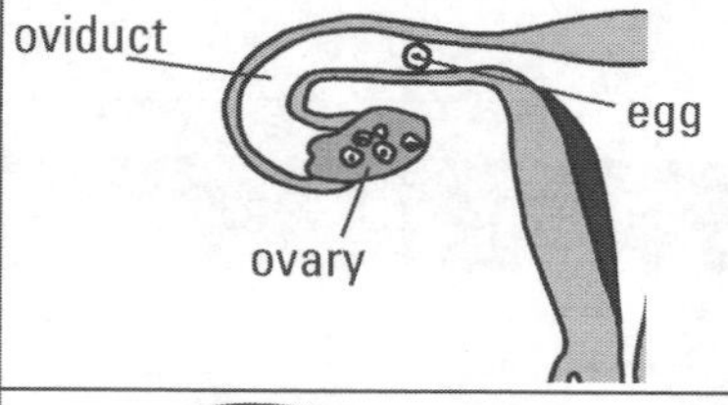 oviduct egg ovary
16–28	The burst follicle now produces the hormone progesterone. It also has an effect on the uterus.	The uterus lining thickens even more in preparation for an embryo.	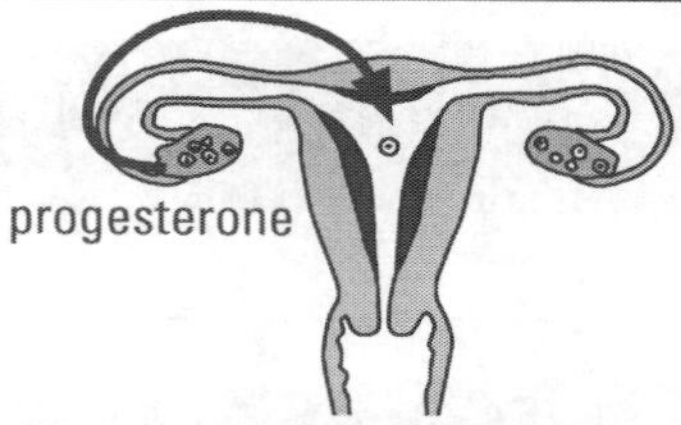 progesterone
also 1–5	Oestrogen and progesterone production stops when fertilisation does not happen and no embryo arrives in the uterus.	The uterus lining breaks down and passes out through the vagina. This is called **menstruation**.	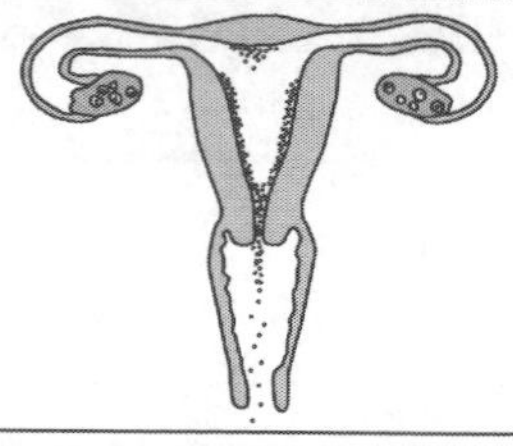

The fluid lost during menstruation is mostly blood. It can be bright red, dark red or dark brown. It sometimes has some clots (dark lumps of blood) in it. Sometimes it seems like a lot of blood, but it is usually less than 100 ml. The blood-forming cells in bone marrow will quickly replace the blood that is lost.

During menstruation some women suffer from mild abdominal cramping. This is because of contractions by the uterus wall as the lining breaks down. This cramping pain is called dysmenorrhoea.

During the days just before menstruation some women suffer from premenstrual syndrome (PMS). The symptoms of this can include feeling depressed, mood swings, tiredness, crying spells, headaches, breast tenderness, changes in appetite and having trouble falling asleep. The exact causes of PMS are not known. They may be linked to the hormone changes that take place in the body during the menstrual cycle. Some symptoms may also be due to a lack of vitamins and minerals in the diet.

During menstruation most women can carry on their normal daily lives. This includes doing PE at school and swimming. The blood that leaves the body during menstruation can be absorbed by wearing a sanitary pad over the opening to the vagina or placing a tampon into the vagina. Menstruation does not take place once a woman is pregnant (see Unit 7). Sometimes a woman stops menstruating for other reasons. These include emotional stress, severe weight loss (due to anorexia, for example) and abnormal hormone levels.

Between 45 and 55 women usually stop menstruating. This time is called menopause.

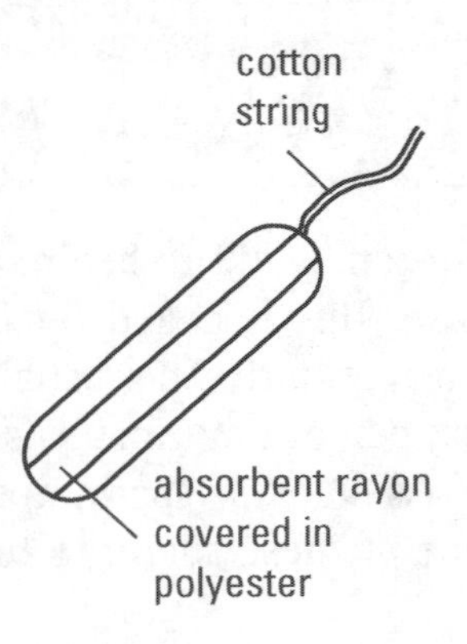

Tampon

4U2DO

1 What happens at the start and end of the menstrual cycle?

______

2 Menstruation

a) How long does a normal period last? ______

b) What happens during this time?

______

c) Why do some women feel cramping pains during their periods?

______

3 For each of the hormones listed below, write down what effect they have on the body.

a) follicle stimulating hormone ______

b) lutenising hormone ______

c) oestrogen ______

d) progesterone ______

4 Ovulation

a) What happens at ovulation?

______

b) When in the cycle does ovulation usually occur? ______

5 What does menopause signal the end of in a woman's life? ______

6 Write down three things you have learned from this unit.

a) ______

b) ______

c) ______

7 Write down anything you need to ask your teacher to explain.

______

UNIT 07 A New Life Begins

For a new life to begin, a male sex cell (sperm) and a female sex cell (egg) must meet and fuse. This is called fertilisation or conception. A woman usually produces only one egg every month. This usually happens around the middle of her menstrual cycle but it can happen at any time so she can get pregnant at any time.

Millions of sperm are released into the vagina during sexual intercourse. It takes only one of these sperm to fuse with the egg for a possible new life to begin.

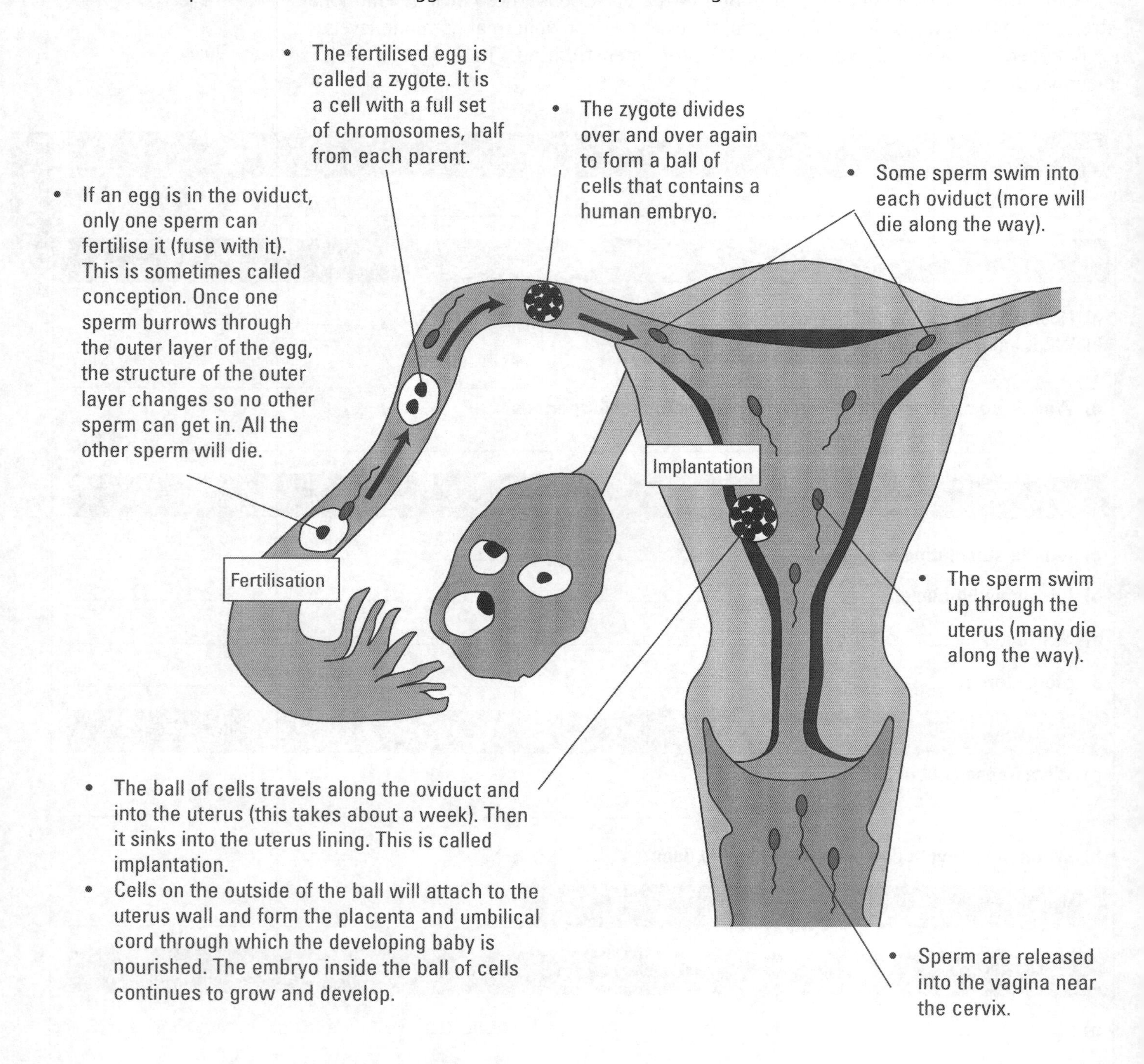

Unless it is fertilised an egg survives for only 24 hours once it has left the ovary. Sperm can survive in oviducts for 48 hours. For fertilisation to take place, intercourse needs to take place near the time of ovulation. While this usually happens around the fourteenth day of the menstrual cycle, it can happen at any time.

Sometimes a woman can produce more than one egg in a month. If each egg is fertilised by a sperm then two babies will develop. These are called fraternal twins. Each baby will have its own placenta. They may be the same sex or they can be different sexes. They may look alike or be quite different to each other. This is because they are formed from two separate eggs and two separate sperm.

Fraternal twins

Identical twins happen when one egg is fertilised by one sperm and the zygote starts to divide as usual but then for some reason (scientists are not sure why) the ball of cells splits into two balls of cells and each develops into a baby. The babies share one placenta. Because they come from only one egg and one sperm, they have the same chromosomes so look identical.

Identical twins

It takes about nine months (38 weeks) from conception to birth. During that time huge changes take place; by week eight the new life growing in the uterus already looks like a baby. It has all its organs and major body structures.

embryo

placenta

After one month the embryo is about 7 mm long. Arm and leg 'buds' have formed, the spine and stomach are beginning to form and the heart is pumping blood to the placenta.

After two months the embryo is about 5 cm long. The arms and legs have grown bigger, the nose, mouth and eyes are starting to form. Organs are developing inside.

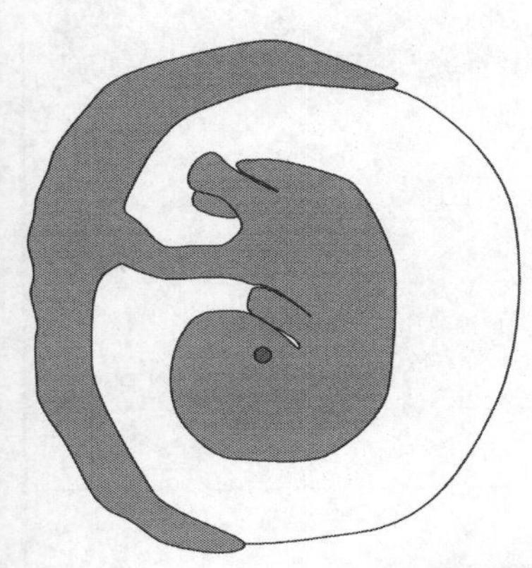

After three months the embryo is called a foetus. All major organs and body structures have formed. It is about 7 cm long, weighs about 28 g.

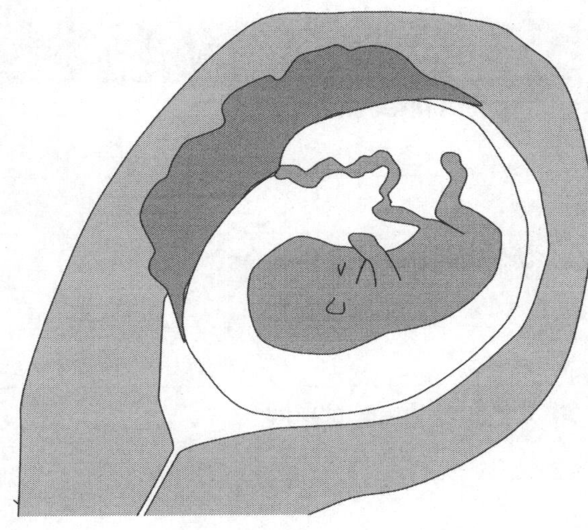

From three to six months the foetus increases in size and its human features become more recognisable. By five months it is 19 cm long and weighs about 500 g. Eyebrows and eyelashes can be seen on the face. It has fingernails and toenails. The whole body is covered in a fine hair called lanugo. The foetus is quite active and the mother can feel it moving. By six months the eyes are open and tooth buds are forming in the gums.

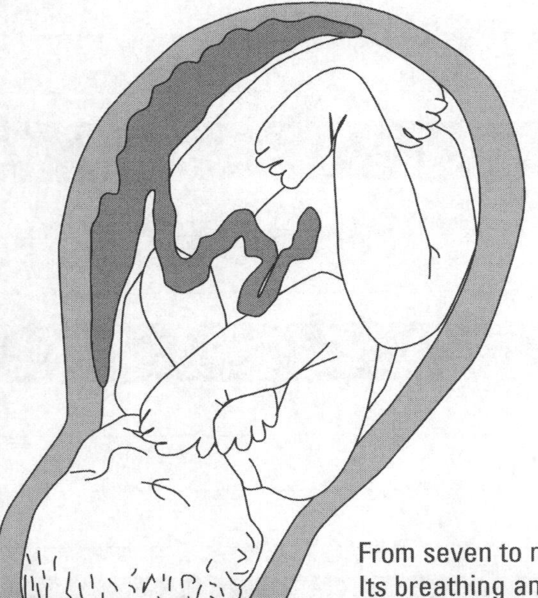

From seven to nine months the baby grows rapidly. Its breathing and circulatory systems get ready to work breathing air and carrying oxygen. Bones begin to harden and muscles thicken. At birth most babies weigh between 2.7 and 4.5 kg and are about 50 cm long.

1 What happens at conception and where does it take place?

2 Sperm

a) Why are so many sperm released into the vagina?

b) How do the sperm move so they can reach the egg?

c) What happens to the sperm that do not fertilise the egg?

3 In what time frame after ovulation does fertilisation need to occur?

4 What does implantation mean?

5 What are the main changes that happen to the baby in

a) the first three months?

b) the second three months?

c) the final three months?

6 Explain the difference between how fraternal and identical twins are made.

7 Write down three things you have learned from this unit.

a)

b)

c)

8 Write down anything you need to ask your teacher to explain.

UNIT 08 BIRTH

During the 38 weeks between conception and birth, a baby needs food and oxygen. It has to get rid of wastes such as carbon dioxide. It must be protected from physical damage as the mother goes about her daily life. Inside the mother's uterus, special structures develop to carry out these tasks.

- **The placenta** This is an organ that grows from the tissue lining the uterus and the outer layer of cells surrounding the embryo. It is soft and spongy. It contains small villi (finger-like folds of tissue) that contain blood vessels from the baby. Small pools of blood from the mother surround these villi. Food and oxygen pass from the mother's blood through the walls of the villi and into the baby's blood. Wastes from the baby move through the villi into the mother's blood. The mother gets rid of these wastes with her own. Blood from the baby does not mix or even touch blood from the mother. Everything that passes between mother and baby has to move through the walls of the villi. At birth, the placenta is a flat 'cake' about 20 cm across and weighing about 450 g.

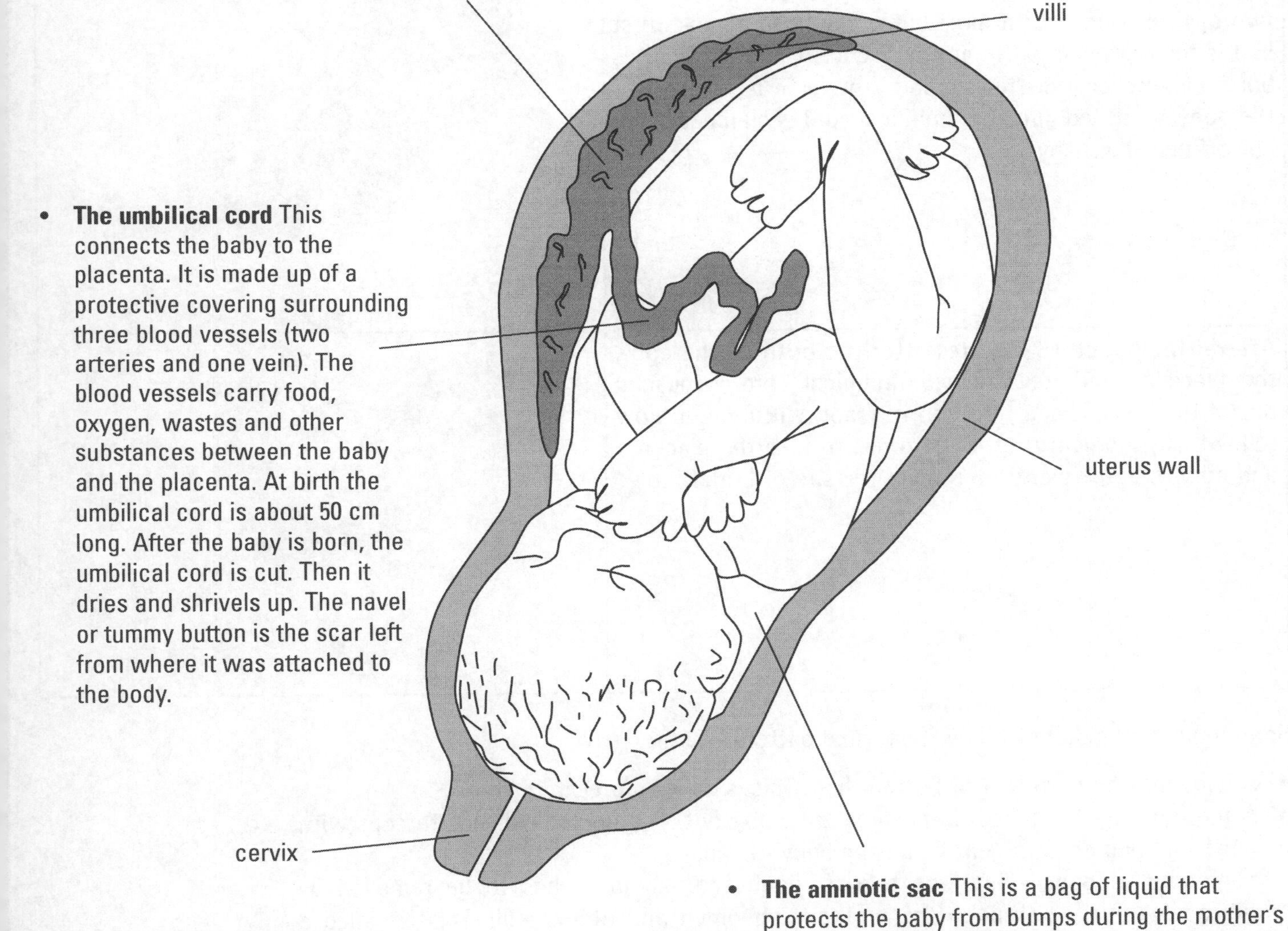

- **The umbilical cord** This connects the baby to the placenta. It is made up of a protective covering surrounding three blood vessels (two arteries and one vein). The blood vessels carry food, oxygen, wastes and other substances between the baby and the placenta. At birth the umbilical cord is about 50 cm long. After the baby is born, the umbilical cord is cut. Then it dries and shrivels up. The navel or tummy button is the scar left from where it was attached to the body.

- **The amniotic sac** This is a bag of liquid that protects the baby from bumps during the mother's everyday activities. It also keeps the baby from drying out. In the first stage of labour before birth, the lining of the sac breaks, releasing the liquid inside. This is when 'the waters have broken'.

After about 38 weeks the baby is ready to survive outside the mother's body. Hormones make the walls of the uterus start to regularly contract and relax. The process of birth begins. There are three stages to this process: labour, delivery and afterbirth.

Stage	
Labour. This is the longest stage. It can last for six to 12 hours or more. The muscular walls of the uterus contract and relax at regular intervals, pushing the baby down towards the cervix. These contractions are called labour pains. The cervix widens until it reaches 10 cm across.	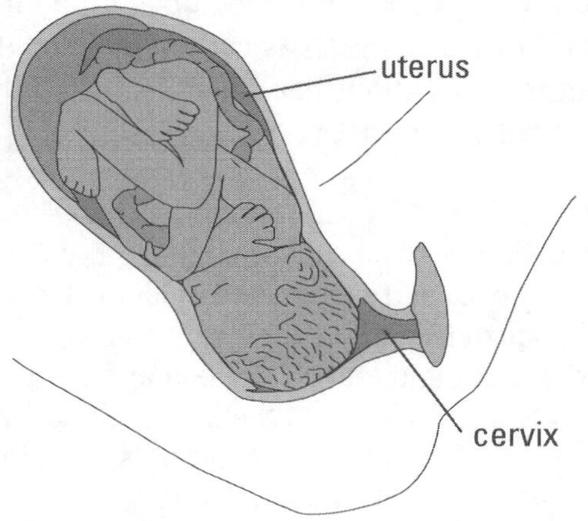 uterus cervix
Delivery. This can take from 20 minutes to several hours. Contractions are now longer and stronger and come every two to three minutes. In most births the head appears first and is facing towards the mother's back. Once the head is born, a midwife or doctor supports the head as the rest of the body is pushed out. The umbilical cord is clamped, then cut off near the baby.	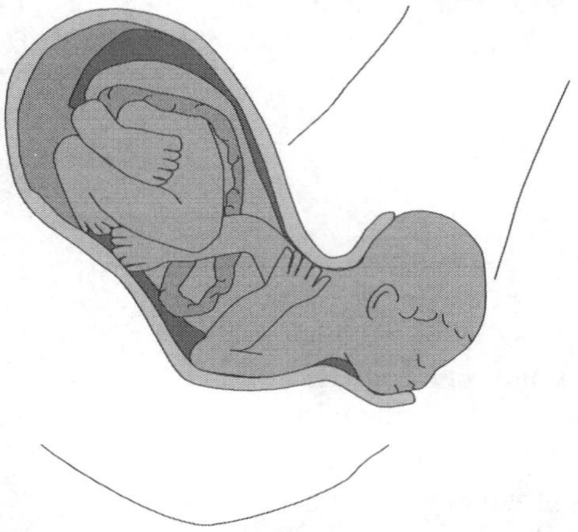
Afterbirth. About 15 minutes after the birth of the baby, the placenta and its attached umbilical cord is pushed out of the body. It is a tradition in Maori culture, and now followed by women of other cultures, to take the placenta and bury it. Some plant a tree over the site of burial.	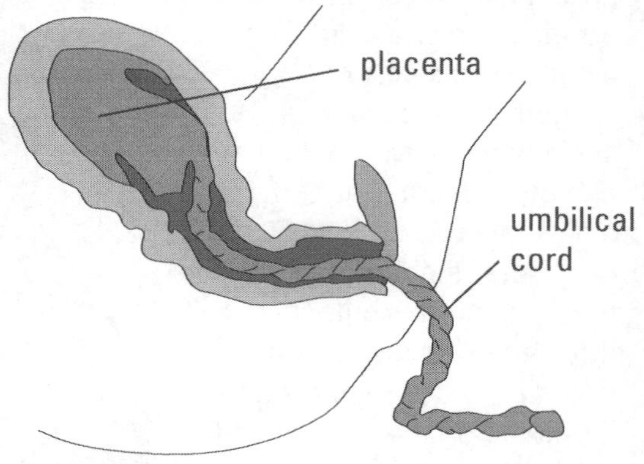 placenta umbilical cord

Sometimes births do not follow this usual pattern.

- A baby may be born feet or bottom first. This is called a breech birth.
- A mother may have trouble pushing the baby out. The doctor will use forceps with large rounded ends to gently pull the baby's head.
- Sometimes a baby cannot be born through the vagina. It has to be removed by the doctor cutting through the mother's abdomen and uterus wall. This is called a caesarean section.

During pregnancy a woman's breasts grow larger and develop the glands to produce milk. The milk produced in a woman's breasts is a complete food for her baby and contains all the food it needs to grow. The milk also has antibodies that protect the baby from disease and infection.

1 Name the main jobs of

a) the placenta

b) the umbilical cord

c) the amniotic sac

2 Not everything that passes across from the mother's blood to the baby are good for the baby's development. What are some things that cross over from the mother and harm a baby as it is developing?

3 Name the three stages of labour and write down one key thing that happens in each stage.

a)

b)

c)

4 Put numbers into the blank circles to show which terms best match the descriptions.

1 villi	◯	bag of liquid
2 placenta	◯	baby not head first
3 navel	◯	cut abdomen and uterus
4 amniotic sac	◯	folds of tissue
5 caesarean	◯	labour pains
6 contractions	◯	soft, spongy organ
7 breech	◯	pull baby out
8 forceps	◯	tummy button

5 Write down three things you have learned from this unit.

a)

b)

c)

6 Write down anything you need to ask your teacher to explain.

Genes and babies

UNIT 09 REPRODUCTIVE TECHNOLOGY

Infertility is not being able to have children. This may be due to physical problems such as

- the man not producing enough sperm
- the sperm not being energetic enough to swim through the vagina and uterus to get to the egg
- the woman not ovulating (releasing eggs)
- the woman's oviducts being blocked, preventing the sperm reaching the egg
- the woman having repeated miscarriages (the birth of a baby before it is able to live).

Assisted reproductive technology involves processes that try to solve these problems and help couples have a child.

Hormone treatment, for example, may increase sperm production and egg production and cause ovulation to occur. Surgery can be used to unblock oviducts. These solutions cannot help everyone though. Sometimes no reason can be found for why couples cannot have children.

In vitro fertilisation (IVF) is a procedure that can help some of these couples. It works like this.

- Eggs are surgically removed from the woman's ovaries.
- The eggs are mixed with sperm in culture dishes and fertilisation occurs.
- About two days after fertilisation, an eight-celled embryo has formed. This embryo is put into the woman's uterus where it may implant into the uterus wall and develop into a baby.

Not all the embryos formed in the culture dish are put into the woman's uterus. Usually two or three are put in because they do not all survive. Putting all the embryos into the mother could result in multiple pregnancies that could prove harmful to both the mother and the babies. Some of the embryos produced by IVF are frozen. They can be used in later treatments. Not all frozen embryos survive the thaw process; only about 70 per cent survive. Those that do survive can be put into the woman's uterus where one or more may develop into a baby.

For a woman who gets pregnant but keeps having miscarriages, a surrogate mother is another means of having a child. The embryo formed from a couple's egg and sperm using in vitro fertilisation is put into the uterus of another woman (the surrogate mother). It implants in her uterus and develops normally. Once the baby is born it is given to the couple that is its biological parents. This is an important social issue, and it raises many ethical, moral and legal questions.

Reproductive technology is not just about helping infertile couples have children. It is also being used to guarantee babies are born without genetic abnormalities. Preimplantation genetic diagnosis (PGD) is an investigative test used in combination with in vitro fertilisation. It is performed on embryos of couples known to have serious inherited genetic conditions that could be passed on to their children. A single cell is removed from an embryo and is analysed for a specific abnormality. Only embryos not affected by the genetic condition are put into the mother's uterus.

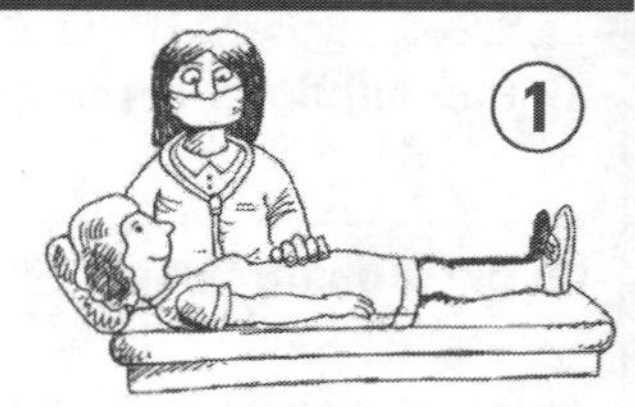

Eggs are removed from the woman. They will be mixed with sperm in culture dishes.

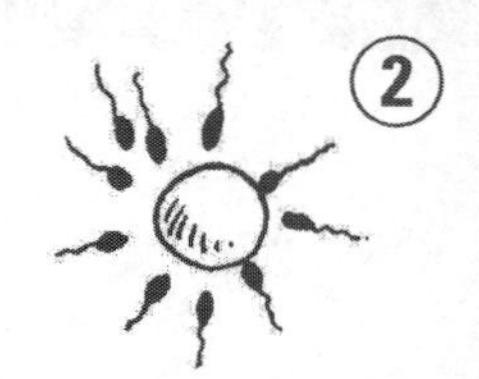

One of the eggs being fertilised by sperm.

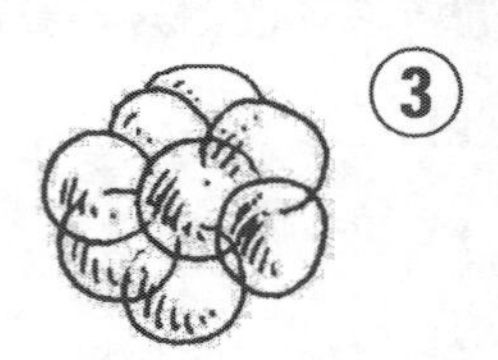

Sperm have fertilised the egg and an embryo has formed.

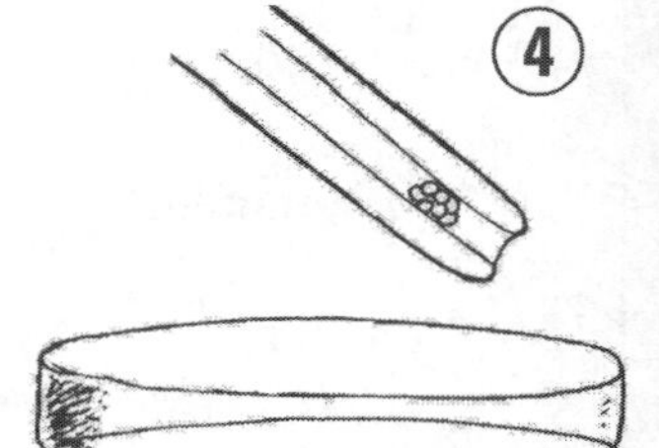

The embryo is collected from the dish and is ready for implantation.

If the embryo implants correctly it will develop into a baby.

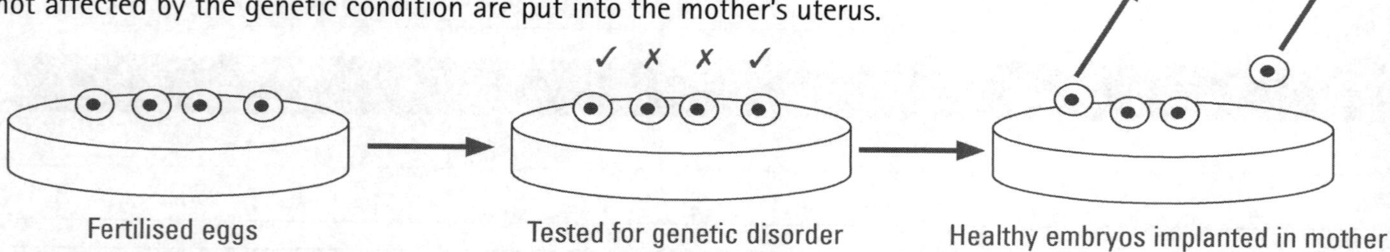

1 What does infertility mean?

2 What are two causes of infertility?

3 What do you think in vitro fertilisation means?

4 Why are not all the embryos produced by in vitro fertilisation put into the woman's uterus?

5 Can you think of some reasons for and against surrogate motherhood?

For	Against

6 Can you think of some reasons for and against the use of preimplantation genetic diagnosis?

For	Against

7 Write down three things you have learned from this unit.

a)

b)

c)

8 Write down anything you need to ask your teacher to explain.

UNIT 10 Sexually Transmitted Infections

Sexually transmitted infections or STIs (once called venereal disease) are spread through sexual contact where body fluids such as semen, blood and vaginal secretions are exchanged with an infected person, or by having sex with an infected person. The bacteria and viruses that cause STIs cannot live for long outside the human body. So you *do not* catch them from toilet seats, sharing a towel or from a sneeze. (Pubic lice can be caught from non-sexual contact.)

People with the greatest risk of getting these infections include those people who

- begin sexual activity at an early age
- have many sexual partners
- have sex with high-risk people (such as an HIV-positive person or an STI-infected person).

Common sexually transmitted infections include chlamydia, genital herpes, pubic lice and HIV.

Chlamydia is caused by a bacterium. It can cause sterility (not being able to reproduce) in both men and women. It is spread through vaginal and anal intercourse and from a mother to her baby in the birth canal. Unfortunately 75 per cent of women and 50 per cent of men get no symptoms so do not know they have chlamydia. If symptoms do appear, it is from seven to 21 days after infection. Symptoms include:

- discharge from the penis or vagina
- excessive vaginal bleeding
- abdominal pain, nausea, fever
- swelling or pain in the testicles.
- pain or burning while urinating, frequent urination
- painful intercourse for women
- inflammation of the rectum or cervix

Chlamydia can be successfully treated with antibiotics. Using condoms reduces the risk of infection.

Two forms of the herpes simplex virus cause genital herpes. The virus remains in the body for life and may cause recurring symptoms for the rest of one's life. It is spread through vaginal, anal and oral sex and sexual intimacy including touching and kissing. Symptoms include:

- a recurring rash with clusters of itchy or painful blistery sores appearing on the vagina, cervix, penis, mouth, anus, buttocks, or elsewhere on the body
- painful ulcerations that occur when blisters break open
- (at the first outbreak) pain and discomfort around the infected area, itching, burning sensations during urination, swollen glands in the groin, fever, headache, and a general run-down feeling.

Symptoms appear two to 20 days after infection. It is most contagious from the time the sores appear until they have healed and the scabs fallen off. There is no cure. Treatment with antiviral drugs may relieve the symptoms and reduce the number of recurrences.

Pubic lice or 'crabs' are small insects that live and lay their eggs in the pubic hair. They are spread through contact with infected bedding, clothing and off toilet seats as well as from sexual contact. Symptoms usually appear about five days after contact and include:

- intense itching in the genitals or anus
- mild fever
- lice or nits (egg sacs) in pubic hair.

Lice can be treated with medication bought from a chemist. All bedding and clothing must be well washed and dried. People who limit their number of sexual partners are at less risk of getting lice.

Human immunodeficiency virus (HIV) weakens the body's immune system making it hard to fight infections. It can cause AIDS (acquired immunodeficiency syndrome). It is the most dangerous of all the sexually transmitted infections and is a leading cause of death around the world. It is spread through blood, semen and vaginal fluids as a result of anal and vaginal intercourse, sharing contaminated needles, contaminated blood transfusions or a simple needle prick. Babies can get it through breast milk. Symptoms may take up to ten years to appear and include:

- constant or rapid unexplained weight loss, diarrhoea, lack of appetite
- fatigue, persistent fevers, night sweats, dry cough
- a thick, whitish coating of yeast on the tongue or mouth – 'thrush'
- severe or recurring vaginal yeast infections
- purplish growths on the skin.

There is no cure or vaccination for HIV but condoms provide protection from the virus. There are some treatments for those suffering from AIDS.

People are often embarrassed and ashamed when they find they have an STI so do not seek treatment. Looking after your sexual health is as important as looking after the rest of you.

The best way to avoid sexually transmitted infections is not to have sex. But because people do have sex, they can reduce the risk of infection by limiting the number of sexual partners and practising safe sex – avoid contact with infected partners, avoid genital sores, and prevent the exchange of body fluids such as semen, blood and vaginal secretions. If you or your partner has any symptoms, see a doctor straight away.

4U2DO

1 STIs

a) What does STI stand for? ______________________________

b) What would your grandparents have called these infections? ______________________________

2 How are STIs spread?

____

____

3 Who is at greatest risk of getting an STI?

____

4 Complete the table.

	Chlamydia	**Genital herpes**	**Pubic lice**	**HIV/Aids**
Cause				
Two symptoms				
Time for symptoms to appear				
Treatment if any				

5 How can a person reduce the risk of getting an STI? ______________________________

____

6 Write down three things you have learned from this unit.

a) ______________________________

b) ______________________________

c) ______________________________

7 Write down anything you need to ask your teacher to explain.

____

UNIT 11 THE ACID FAMILY

Acids are a family of chemicals that have many features or properties in common. Some of these properties are physical (those that do not involve chemical reactions) while others are chemical (how a substance reacts with other chemicals). From your own experiences you may recognise these properties:

a) Acids have a sour or sharp taste. The word acid comes from the Latin word *acidus*, which means sour. (*Never* taste lab acids.)

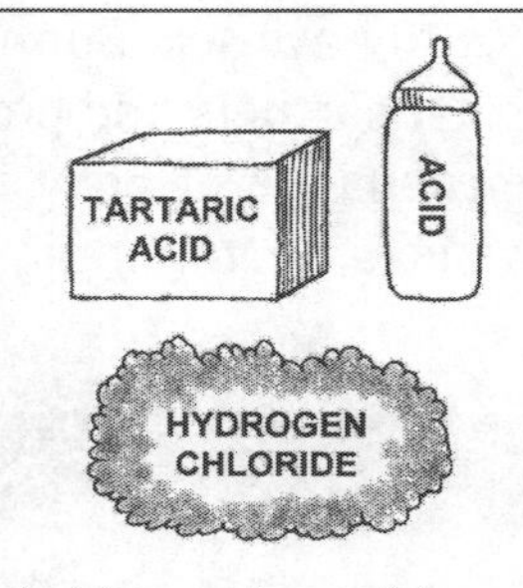

b) Acids can be solids, liquids or gases. They show their acidic properties only when they are dissolved in water.

c) Acids are corrosive – they burn or irritate your skin and eat away at other substances.

d) Acids react with carbonates and make carbon dioxide gas.

One group of acids are the organic acids. They can be found in plants and animals and also in food and drink. The organic acids include:

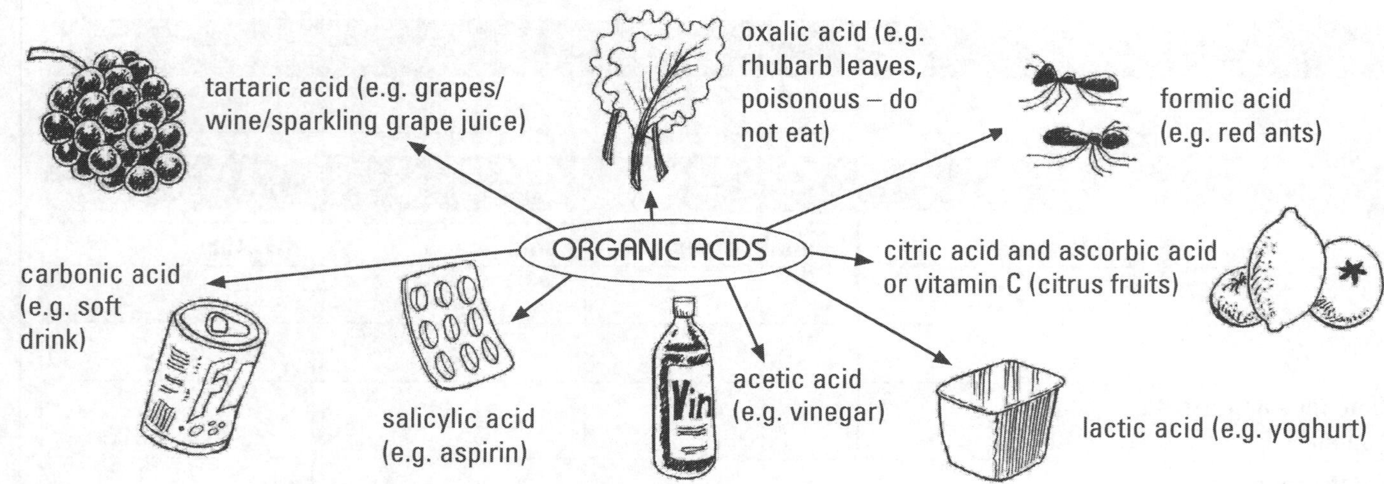

Another group of acids are the inorganic or mineral acids. These acids are made from the minerals found in rocks and in the Earth's crust. They include the hydrochloric acid, sulfuric acid and nitric acid you use in experiments in a school science lab. These acids are stronger than organic acids so need to be handled carefully.

Acids are made of atoms. Another feature acids have in common can be seen in their chemical formulas.

Acid	Chemical formula
acetic	CH_3COOH
citric	$C_5H_7O_5COOH$
tartaric	$C_3H_5O_4COOH$
hydrochloric	HCl
nitric	HNO_3
sulfuric	H_2SO_4

All acids contain hydrogen (H), and the organic acids all contain the element carbon (C) as well.

Scientists say an acid is a substance that releases hydrogen ions (H^+) when it is dissolved in water. It is these hydrogen ions that make a solution acidic.

Strong acids like hydrochloric and sulfuric release many hydrogen ions when they dissolve in water.

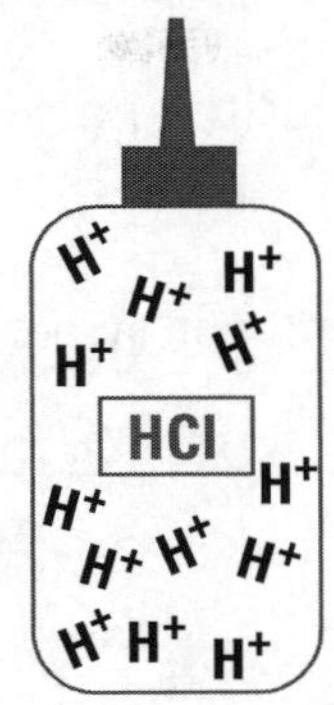

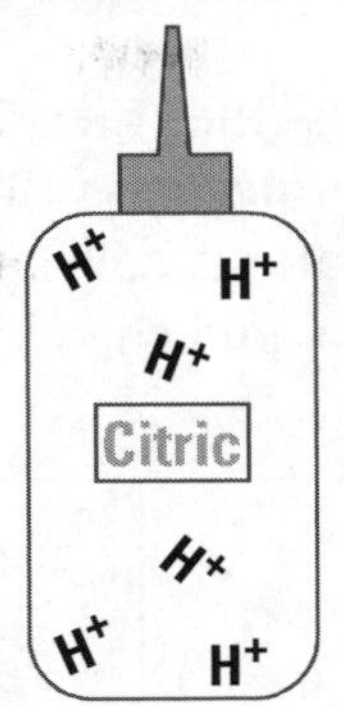

Weak acids like acetic and citric acid release only a few hydrogen ions when they dissolve in water.

4U2DO

1 Acid properties

a) Why are acidic compounds called a family of chemicals?

____

b) Describe two physical properties of acids.

____

c) Describe one chemical property of acids.

____

d) Why should you never sniff or taste acids?

____

2 Acid groups

a) Write the name and chemical formula for two mineral acids you will use in the lab.

____

b) Circle the organic acids in this list.

sulfuric lactic acetic hydrochloric tartaric nitric

3 Acids and water

a) All acids contain atoms of which chemical element?

____

b) When hydrogen chloride crystals are put into water, a solution of hydrochloric acid is made. What makes the solution acidic?

____

c) Why is vinegar a weak acid?

____

4 Write down three things you have learned from this unit.

a) ______________________________

b) ______________________________

c) ______________________________

5 Write down anything you need to ask your teacher to explain.

____

Acids and bases

UNIT 12 THE BASE FAMILY

Bases are another family of chemicals. They are said to be the opposite of acids because they react with acids to neutralise them or decrease their acidity.

A base that can be dissolved in water is called an alkali.

Bases share some common physical and chemical features. From your own experiences you may recognise these features:

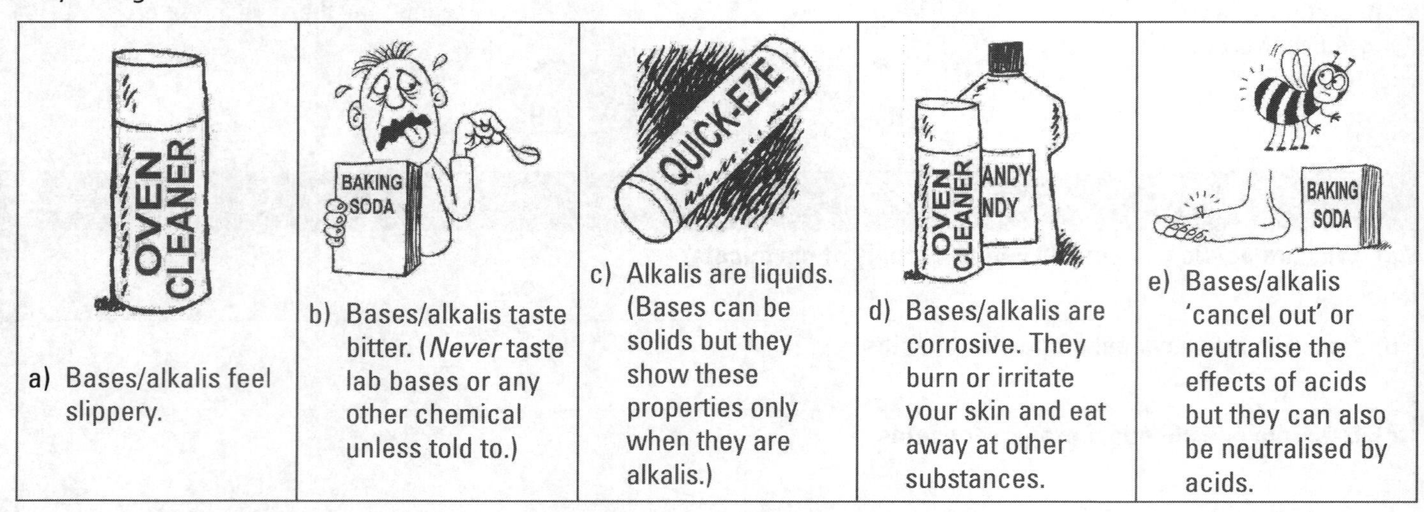

a) Bases/alkalis feel slippery.

b) Bases/alkalis taste bitter. (*Never* taste lab bases or any other chemical unless told to.)

c) Alkalis are liquids. (Bases can be solids but they show these properties only when they are alkalis.)

d) Bases/alkalis are corrosive. They burn or irritate your skin and eat away at other substances.

e) Bases/alkalis 'cancel out' or neutralise the effects of acids but they can also be neutralised by acids.

Bases and alkalis are found around the home as well as in the science lab. For example:

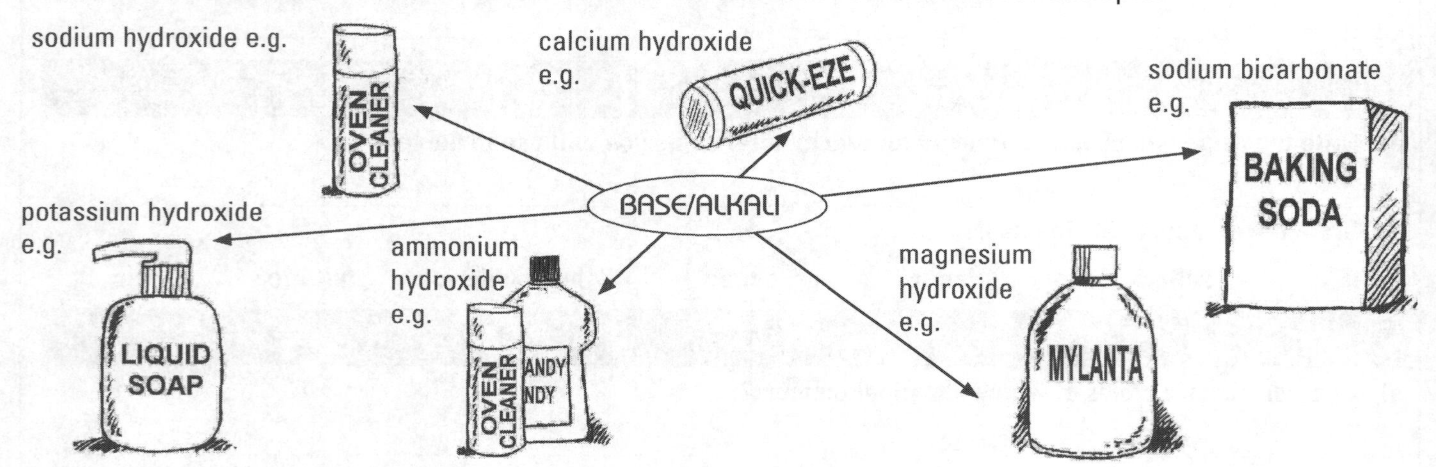

Many of these bases are hydroxides. This means they contain hydroxide ions (OH^-).

Base	Common name	Chemical formula
sodium hydroxide	caustic soda or lye	$NaOH$
potassium hydroxide	caustic potash	KOH
calcium hydroxide	slaked lime	$Ca(OH)_2$
magnesium hydroxide	milk of magnesia	$Mg(OH)_2$

Strong bases, such as sodium hydroxide, release many hydroxide ions when they are dissolved in water. They form strong alkaline solutions. Weak bases, such as magnesium hydroxide, release only a few hydroxide ions to form weak alkaline solutions.

When a base is added to an acid, the hydroxide ions from the base combine with the hydrogen ions from the acid to form water. This reaction neutralises the acid.

$$H^+ + OH^- \longrightarrow H_2O$$

Bases and alkalis can be just as dangerous as acids. Caustic soda or lye is used in oven cleaners (for example Mr Muscle) and drain cleaners (for example Drano). It is very corrosive, especially in a concentrated form. When using both these chemicals, it is advisable to wear gloves and eye protection and to avoid breathing in the fumes.

It is dangerous to mix different types of household cleaners. For example, cleaners that contain ammonium hydroxide, such as Handy Andy, should never be mixed with bleaches containing sodium hypochlorite, such as Janola. Mixing these two cleaners produces a chemical reaction that sets off a very poisonous gas called chloramine.

4U2DO

1 Bases

a) Why are the bases called a family of chemicals?

b) Describe two physical properties of bases.

c) Describe one chemical property of bases.

d) Explain the difference between a base and an alkali.

2 Bases can neutralise an acid. Draw a chemical equation to show what is happening in a neutralisation reaction.

3 In the spaces below, draw diagrams to show the difference between a strong alkali and a weak alkali.

4 Caustic

a) Find out what the word 'caustic' means in the name 'caustic soda' and write it here.

b) Why should you wear protective gloves and have the windows open when you use a caustic oven cleaner?

c) Why should you never mix household cleaners?

5 Write down three things you have learned from this unit.

a)

b)

c)

6 Write down anything you need to ask your teacher to explain.

UNIT 13 Chemical Indicators

Litmus dye is extracted from lichen

Chemical indicators are made from dyes extracted from plants. Some indicators just show whether a substance is an acid or an alkali. Others show the strength of the acid or alkali.

Litmus is a dye that dissolves in water. It is extracted from some types of lichen (pronounced *lie-k'n*). Soaking absorbent paper in litmus dye and then drying it makes litmus paper, which is used to see if a solution is an acid or a base.

Blue litmus paper turns red when dipped in an acid (remember acid = red) but stays blue in an alkali. Red litmus paper turns blue in an alkali (remember base = blue) but stays red in an acid.

Universal indicator is a mixture of several indicators. It can be a green-coloured liquid you add to a solution or a pale yellow-orange-coloured paper strip you dip into a solution. The colour the universal indicator turns when it is put into a solution can be used to indicate the solution's pH, for example red = pH 1, green = pH 7, purple = pH 14.

Colour of universal indicator	Strength of acid
red	strong acid
orange	medium acid
yellow	weak acid
green	neutral
pale blue	weak base
pale purple	medium base
dark purple	strong base

The pH scale is a more accurate way of measuring the acidity of a solution. It measures the quantity of hydrogen ions present in a solution. The scale goes from 0 to 14.

0		7		14
strong acid	getting weaker →	neutral	← getting weaker	strong base

pH	Example	pH	Example
0	lab acids, e.g. HCl, H_2SO_4	8	baking soda (8.4)
1	battery acid	9	all-purpose cleaner
2	lemon juice, stomach acid	10	Mylanta
3	vinegar (3.5)	11	ammonia
4	tomato (4.5)	12	Janola bleach
5	cabbage (5.2–5.4)	13	oven cleaner
6	milk (6.2)	14	lab bases, e.g. NaOH
7	distilled water		

pH paper changes colour like universal indicator paper when it is dipped in acid and alkaline solutions. There are also digital pH meters, which have a probe that is dipped into the solution to be tested and shows a digital reading of the pH.

Many plants are natural indicators of soil pH. Hydrangeas, a common flowering shrub grown in New Zealand, have different petal colours depending on the acidity of the soil.

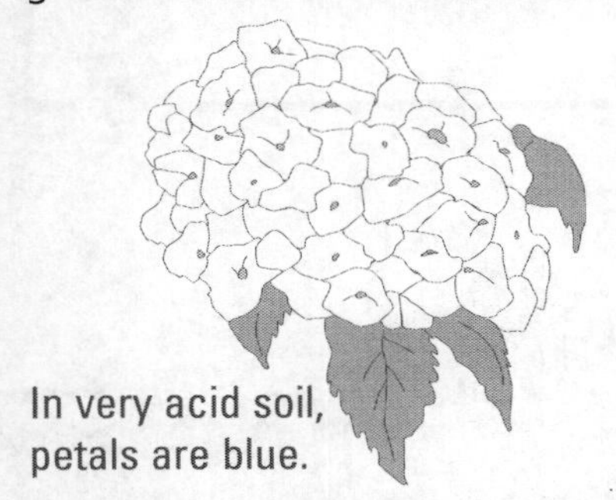

In very acid soil, petals are blue.

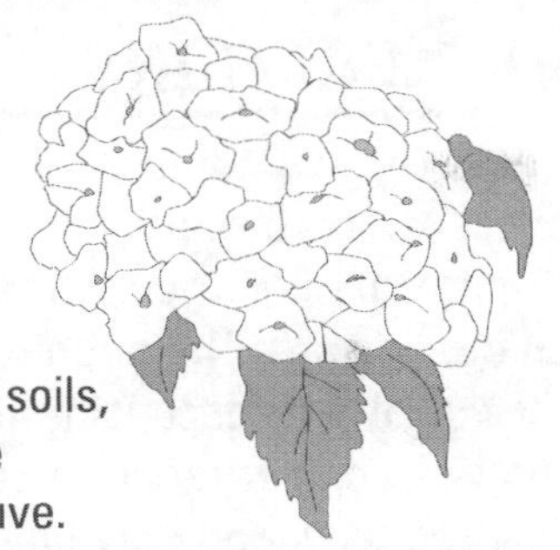

In less acidic soils, the petals are purple or mauve.

In neutral soils, the petals are red or pink.

4U2DO

1 Indicators

a) What are chemical indicators used for?

b) What are they made from?

c) Name two different indicators you could use to show whether a solution was acidic or basic.

2 pH scale

a) What does the pH scale measure?

b) Put the following substances in the correct place on the pH scale drawn below: blood (pH 7.4), shampoo (7), yoghurt (4), Janola (12), wine (3), seawater (8.4), limewater (11), tea (5), toothpaste (8.5).

0 ________________ 7 ________________ 14

c) Shade the acid side of the scale red, the base side blue, and the neutral pH green.

3 Work out answers

a) Some people have a slice of lemon in their tea instead of milk. Tea with lemon in is a much paler colour than tea without lemon. Explain why.

b) If a hydrangea shrub has pink petals, what does it tell you about the soil?

4 Write down three things you have learned from this unit.

a)

b)

c)

5 Write down anything you need to ask your teacher to explain.

UNIT 14 ACID-CARBONATE REACTIONS

Baking powder is used in cooking to make cakes and biscuits rise. A chemical reaction between two ingredients cream of tartar (tartaric acid) and baking soda (sodium bicarbonate) produces carbon dioxide gas. The bubbles of carbon dioxide get trapped in the cake batter or biscuit mixture, causing it to rise.

A similar reaction occurs when vinegar (acetic acid) is added to baking soda. The mixture bubbles up as carbon dioxide is produced. You may have used this reaction in a science fair project to show a volcano erupting. This reaction can also be used to unblock kitchen pipes. Pour half a cup of baking soda down the plughole, then pour in some vinegar and quickly put in the plug. The carbon dioxide produced in the reaction puts pressure on the blockage and can push it out of the pipe.

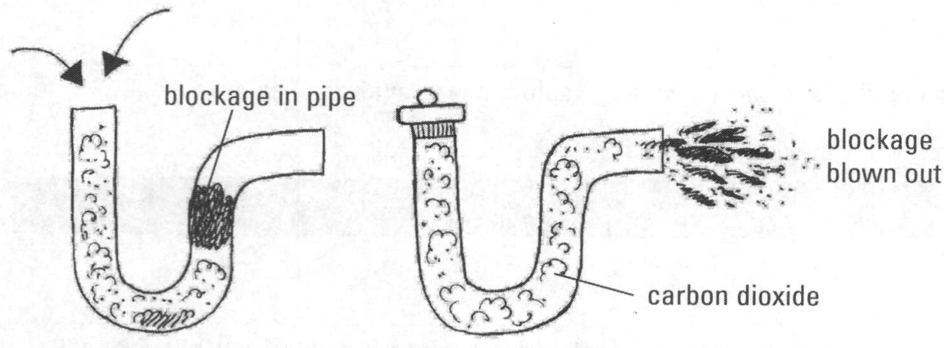

Sometimes people eat or drink too much and they feel 'bloated' and uncomfortable. They say they have indigestion. When this happens they may take health salts such as Eno or Andrews. These powders contain citric or tartaric acid and baking soda. When mixed with water they produce a fizzy drink. In your stomach the carbon dioxide produced makes you burp and this removes the trapped air in the stomach that has been causing the bloated, uncomfortable feeling.

Quick-eze tablets contain calcium carbonate and Gaviscon tablets contain sodium bicarbonate. Both of these bases react with stomach acid to produce carbon dioxide. A similar reaction occurs when you use toothpastes containing calcium carbonate or baking soda. These ingredients help neutralise the plaque acid that the bacteria on your teeth produce.

There is a pattern in these reactions between acids and carbonate compounds. The chemical reaction can be summarised as:

An acid plus a carbonate ⟶ a 'salt' plus carbon dioxide plus water

The reaction that takes place in your stomach when you use Quick-eze can be seen in the science lab when hydrochloric acid is added to calcium carbonate (marble chips). The chemical equation for this reaction would be:

hydrochloric acid	+	calcium carbonate	⟶	calcium chloride salt	+	carbon dioxide	+	water
$2HCl$	+	$CaCO_3$	⟶	$CaCl_2$	+	CO_2	+	H_2O

When the gas produced in this reaction is bubbled through limewater, the limewater turns milky or cloudy. This shows that the gas is carbon dioxide.

The water and 'salt' produced in the reaction are harmless to the body and can be removed from the blood by the kidneys and passed out of the body in urine.

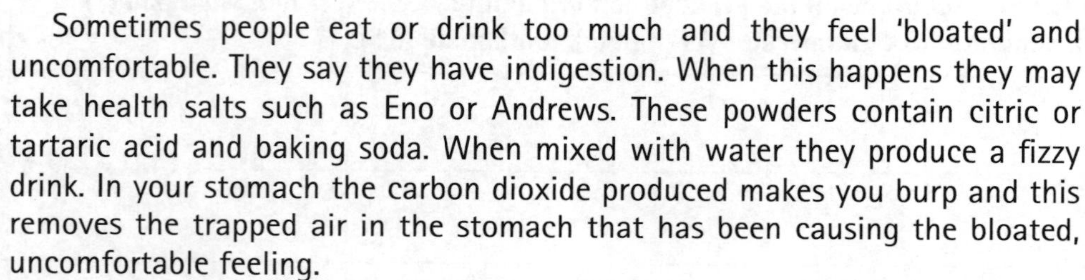

When carbon dioxide in the atmosphere dissolves in water droplets, it forms weak carbonic acid (pH 6). These droplets make up clouds and rain and this means rain is naturally slightly acidic. As this rain falls on buildings and monuments made of limestone (calcium carbonate), it slowly eats them away. It also makes the ground it lands on slightly acidic. The term acid rain refers to rain that is more acidic than normal owing to acidic gases such as sulfur dioxide being released into the atmosphere.

Rain falling on rocks causes chemical weathering. The salt produced in this reaction (calcium bicarbonate) then dissolves in rain and is carried away. This can create limestone caves such as those at Waitomo in the central North Island and the Ngarua Caves in Takaka Hill.

4U2DO

1 Work out answers

a) Name the gas produced when acids react with carbonates.

____

b) You observe fizzing during a reaction in science. What would you do to prove the gas was carbon dioxide and what result would you expect to see?

____

____

c) Explain why baking powder makes cakes rise.

____

____

2 Equations

a) Complete this general equation for acids and carbonates:

acid + carbonate ⟶ ____________ + ____________ + ____________

b) Now complete these word equations:

(i) calcium carbonate plus hydrochloric acid ______________________________

(ii) magnesium carbonate plus sulfuric acid ______________________________

3 Why do indigestion remedies such as Quick-eze and Gaviscon contain carbonates?

____

4 Write down three things you have learned from this unit.

a) ______________________________

b) ______________________________

c) ______________________________

5 Write down anything you need to ask your teacher to explain.

____

UNIT 15 ACID-BASE REACTIONS

Acids and alkalis are said to be opposites. If you mix the right amount of acid with an alkali, then they will cancel each other out and produce a solution that is neither acid nor alkali. The solution is said to be neutral.

People who suffer from acid indigestion and heartburn and those who suffer pain from gastric ulcers (open sores in the wall of the stomach irritated by stomach acid) may take antacids to stop the pain.

Antacids are a group of chemicals that neutralise stomach acid. Mylanta, a liquid antacid, contains the base magnesium hydroxide and Gaviscon antacid tablets contain aluminium hydroxide. Both of these bases neutralise stomach acid. When a base such as magnesium hydroxide reacts with an acid to reduce the acid's acidity, it is called a neutralisation reaction.

The products of the reaction are water and a 'salt'. The reaction can be summarised as:

An acid plus a base ⟶ a 'salt' plus water

When Mylanta neutralises stomach acid the reaction is:

magnesium hydroxide + hydrochloric acid ⟶ magnesium chloride salt + water

$$Mg(OH)_2 + 2HCl \longrightarrow MgCl_2 + 2H_2O$$

When a honeybee stings you, it injects venom that causes pain, swelling, redness and itching. One common treatment of bee stings is to scrape the sting out of the skin then put damp baking soda on the area. Bee venom is acidic (pH 5.5) and is neutralised by the sodium bicarbonate base. Sodium bicarbonate also draws fluid out and reduces itching and swelling. (Remember baking soda for bee.)

Bee

When a wasp stings it does not leave its sting in your skin but the venom it injects produces the same pain, swelling and redness that a bee sting does. Wasp venom is alkaline, not acidic, so a common remedy is to put vinegar (acetic acid) or lemon juice (citric acid) on the site of the sting. The acid acts to neutralise the base. (Remember VW – vinegar for wasp.)

Large tankers on the roads transporting corrosive materials such as sulfuric acid and sodium hydroxide are occasionally involved in crashes. If the tanker is ruptured and the contents leak onto the road, they can be neutralised by using the opposite 'family' of chemicals. Specialists in this kind of accident know exactly what to use. Then lots of water is used to dilute the product, making it less harmful.

If you spill acid on yourself in a science lab, immediately rinse off the area with plenty of water. *Do not* try to neutralise it using a base or alkali because too much of these can be just as damaging to your skin as the acid you are trying to get rid of.

Wasp

4U2DO

1 Antacids

a) What are antacids?

____

b) Write the name and formula of the stomach acid that antacids affect.

____

4U2DO

2 Equations

a) Complete this general equation for acids and bases:

acid + base ——→ ____________ + ____________

b) Now complete these word equations:

(i) hydrochloric acid + sodium hydroxide ____________

(ii) sulfuric acid + magnesium hydroxide ____________

3 When an antacid neutralises the stomach acid, what do you think happens to the products of the chemical reaction?

____

4 Can you write a rhyme or mnemonic to help you remember that you use baking soda on bee stings and vinegar on wasp stings?

____

____

5 Spilled acid

a) If you spill acid on your skin during an experiment, what is the first thing you should do?

____

b) What should you not do and why not?

____

6 Use the clues to complete the puzzle and find the word that goes down.

chemicals that neutralise stomach acid

use baking soda if one of these stings you

this type of solution is neither an acid nor a base

one of the products when acids and bases react

the type of acid found in the stomach

use vinegar if one of these stings you

the other product of an acid-base reaction

causes heartburn and indigestion

opposite of an acid

what you use on a wasp sting

The 'down' word is ____________

7 Write down three things you have learned from this unit.

a) ____________

b) ____________

c) ____________

8 Write down anything you need to ask your teacher to explain.

____

UNIT 16 ACIDS AND METALS

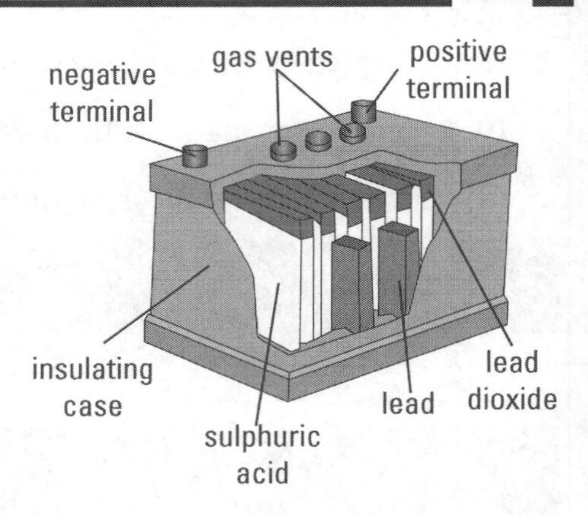

Acid-metal reactions take place in car batteries to produce enough electric current to start the engine. Car batteries are made up of six two-volt cells. Each cell consists of one lead electrode or plate and one lead dioxide electrode. These electrodes are immersed in sulfuric acid. Reactions between the acid and the electrodes produce the electrons that make up an electric current. Vent holes in the battery case allow the small amount of hydrogen gas made during the reaction to escape. Some of the gas is recombined with oxygen in the battery to produce water.

In the science lab, when a piece of magnesium ribbon is put into a test tube of hydrochloric acid, bubbles form on the magnesium. As the reaction continues, the magnesium disappears. Fizzing is a sign that a gas is being produced. Acids contain hydrogen so it is likely the gas is hydrogen gas. If the gas is collected and then released next to a lit match, it explodes with a loud 'pop', proving it is hydrogen.

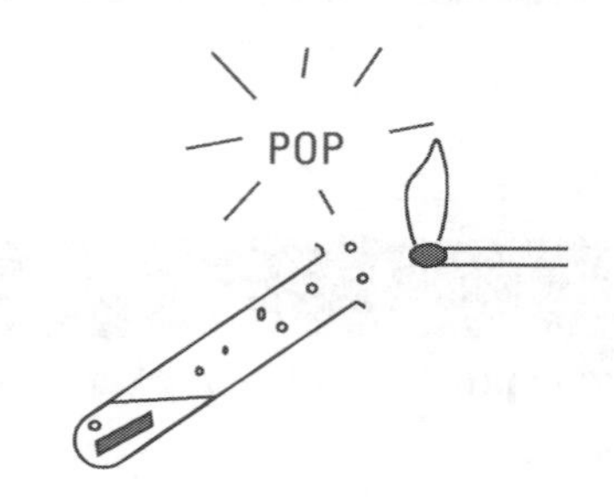

The reaction can be summarised as:

An acid plus a metal ⟶ a 'salt' plus hydrogen gas

For example:

magnesium	+	hydrochloric acid	⟶	magnesium chloride	+	hydrogen gas
Mg	+	2HCl	⟶	$MgCl_2$	+	H_2

Acid rain (discussed in Unit 14) can also corrode metal structures such as bridges and statues as well as cars and outdoor furniture.

Acids can be used to eat away or 'etch' metals in a form of artwork. Metal plates of copper, zinc and iron are used.

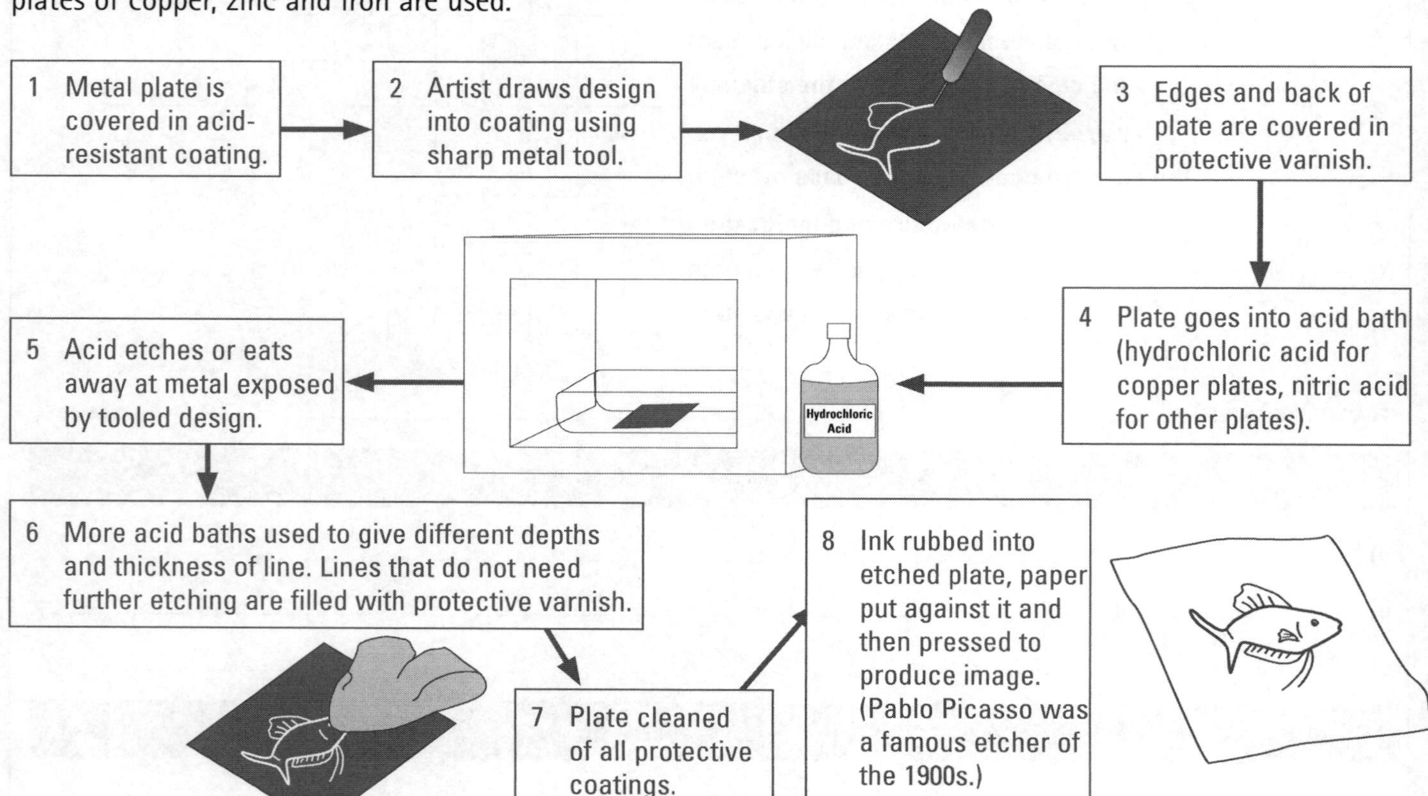

4U2DO

1 A student put a piece of magnesium ribbon into a test tube of hydrochloric acid. A fizzing reaction occurred.

a) What does this observation mean?

b) Explain how the student showed what gas is being produced.

c) Write a word equation for this reaction.

2 Acid rain affects iron bridges and statues. Why does the effect take longer to be seen than the reaction in the test tube in question 1?

3 Battery

a) Write a word equation to describe the reaction that takes place in a lead-acid battery.

b) Recombining hydrogen gas, produced in the battery, to form water is good for two reasons. What are they?

4 Draw a design you could use to produce a piece of artwork from using the etching technique described above.

5 Write down three things you have learned from this unit.

a)

b)

c)

6 Write down anything you need to ask your teacher to explain.

UNIT 17 WHAT IS A SALT?

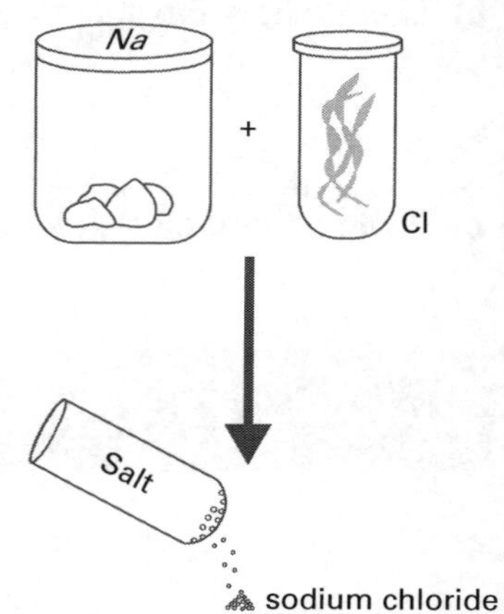

In the previous units you will have noticed that one product of the chemical reactions involving acids was a salt. Most people, when they hear the word salt, think of the white crystals they sprinkle on their food. The salt we put on food is a salt called sodium chloride, but there are many other compounds in the family of 'salts'.

A salt is a compound formed in a reaction between an acid and a base when a metal or another positive ion replaces the hydrogen in the acid.

Salts have two parts to their name. This table shows you how to name a salt.

First name	Second name
Comes from the metal present in the reactants that take part in the reaction.	This name depends on which acid is used in the reaction. It is the part of the acid that is not hydrogen and is easiest to work out if you look at the acid's formula.
For example: sodium, calcium, aluminium and magnesium	For example: hydrochloric acid (HCl) makes *chlorides* (hydrochloric acid is hydrogen chloride dissolved in water) sulfuric acid (H_2SO_4) makes *sulfates* (sulfuric acid is hydrogen sulfate dissolved in water)

Salts are crystals that contain positive and negative ions locked together in a regular pattern called an ionic lattice. When the salt is dissolved in water, the ions are released from the lattice and move randomly about in solution.

When a chemical reaction has finished, the salt is usually dissolved in water so it cannot be seen. For you to see the salt, the water must be evaporated first. Evaporation is the way table salt (sodium chloride) is collected from seawater at Lake Grassmere near Blenheim.

Sodium chloride is the most common salt in seawater but it is not the only salt there. Salts make up about 3.5 per cent of seawater. The most common ions in seawater are chloride, sodium, sulfate, magnesium, calcium and potassium. These ions were once salts contained in the earth's crust.

When rocks are broken down (weathered) and then transported by rivers to the sea, the salt ions add to the saltiness (salinity) of the sea.

The Dead Sea, in the Middle East (Israel/Jordan), has the saltiest water on earth. The water is nine times saltier than the ocean. It contains sodium chloride, calcium chloride and potassium chloride.

Salts are used in many industries and in agriculture. For example:

Ammonium nitrate is used in the production of fertilisers and returns nitrogen to the soil. Plants and animals use nitrogen to make protein. 	Potassium carbonate (potash) is used to make some types of glass and soaps. 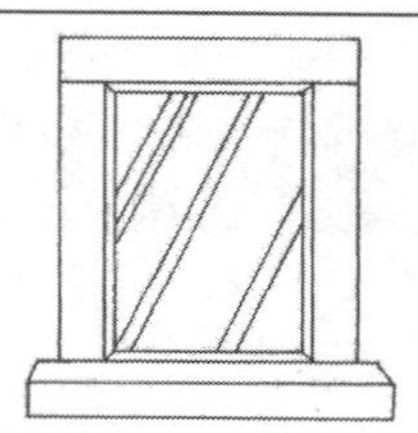
Calcium sulfate (gypsum), once heated, forms plaster of Paris. It is used to make plaster casts used to immobilise broken bones and in the building industry to make wallboard. 	Sodium chloride is used to de-ice roads in winter, as a food preservative, in salt licks for cattle and horses, and as a source of chlorine for use in industry. 

4U2DO

1 In your own words explain what a salt is.

2 Which acids have been used to make the following salts?

sodium chloride ______

calcium sulfate ______

ammonium nitrate ______

3 Equations

a) Complete the following word equations:

(i) **magnesium +** ______ → ______ **chloride** + hydrogen gas

(ii) **calcium carbonate +** ______ → ______ **sulfate** + water + carbon dioxide

b) (for experts) Use chemical formulas to write a chemical equation for the reaction shown in 3 a) (i).

4 Including equipment available in the school science lab, draw and label a diagram showing how you would get the salt out of the water it is dissolved in.

5 Describe two ways salts are used in our everyday lives.

6 Write down three things you have learned from this unit.

a) ______

b) ______

c) ______

7 Write down anything you need to ask your teacher to explain.

UNIT 18 ACIDS, BASES AND THE HUMAN BODY

Glands in the dermis layer of the skin produce sweat (salty and acidic) while sebaceous glands produce oil called sebum. Sweat and oil combine to produce a protective layer on the skin. The acidity (pH 4–5.5) protects you from infection by keeping a control on the number of bacteria that live on your skin.

Plaque is a sticky film of saliva, food particles and bacteria found on the teeth. When the bacteria digest sugary food particles, they produce an acid that can dissolve tooth enamel. Saliva has a neutralising effect on this acid because it is slightly basic (pH 7.4).

Sometimes during strenuous exercise you can get a 'stitch'. This causes muscle tiredness and pain and is due to a build-up of lactic acid in your muscle cells. Lactic acid builds up when cells are trying to make the energy they need to work but are not getting enough oxygen to do so by normal aerobic respiration. When you stop exercising, your blood carries the lactic acid away from the cells and the pain goes away.

DNA or deoxyribonucleic acid is the building block of all living things. Our chromosomes are made up of DNA. Small sections of chromosomes, called genes, carry the instructions about what we look like and how our body will grow and function.

Amino acids are the building blocks of protein. Proteins are used to build and repair tissue. Every cell in our body contains protein. Enzymes are special types of proteins that control the chemical reactions that take place in our bodies.

Blood has a pH of 7.4, which means it is slightly basic. When carbon dioxide passes from cells into the blood, it dissolves in the plasma to form carbonic acid (H_2CO_3). This is then broken down to form bicarbonate ions (HCO_3) and these reduce the acidity of the blood.

The pH of urine varies over the day. It is more acidic in the morning (pH 6.5–7) but becomes more alkaline by evening (pH 7.5–8).

Glands in the stomach wall secrete hydrochloric acid (pH 2) into the stomach. One reason for this is because the enzyme pepsin, which breaks down protein, works best in an acidic environment. This acid also kills any microorganisms in the food that could make you sick.

When you vomit, it is the acid that burns your throat and it can attack your tooth enamel in the same way that plaque acid does. People who suffer from bulimia (an eating disorder characterised by binge eating and then vomiting) often have damaged teeth for this reason.

Sometimes too much stomach acid is produced. It can eat away at the stomach lining causing painful ulcers. Taking antacid tablets can help soothe the pain.

The B-complex vitamin folic acid is present in fresh green vegetables and in cereals and cereal products. Unfortunately cooking and food processing destroy 50–90 per cent of folic acid in food so it is added to some foods such as pasta and rice. It is believed folic acid prevents birth defects such as spina bifida, where the spinal cord does not form properly.

ANSWERS (this 4-page section is removable)

TOPIC 1 GENES AND BABIES

Unit 1 HUMAN VARIATION

1 individual
2 individual
3 individual
4 a) individual e.g. height, hair colour; b) colours blend into one another; hair is not one colour or another
5 a) individual e.g. German shepherd; b) individual e.g. Granny Smith
6 for survival of species

Unit 2 CHROMOSOMES, GENES AND DNA

1 1st diagram = nucleus, 2nd = chromosome and genes, 3rd = DNA and bases
2 a) chapter, b) paragraph, c) sentence, d) word
3 have the same biological parents or get genes from the same parents
4 a) they get it from their mother who only has X sex chromosomes to pass on; b) father passes on the Y chromosome; c) 50%

Unit 3 PUBERTY

1 it produces hormones
2 a) chemical messengers; b) in the blood
3 because a baby is born with the primary characteristics that make them a boy or a girl
4 girls only – e.g. breasts get larger, hips widen, menstruation starts; girls and boys – get taller, produce sex hormones, pubic hair grows; boys only – sperm is produced, voice deepens, facial hair grows
5 individual e.g. people are different and have different chromosomes
6 individual e.g. emotionally and socially

Unit 4 THE MALE REPRODUCTIVE SYSTEM

1 produce sperm, produce male hormones, get sperm into woman's body
2 blood filling the spongy tissue of the penis
3 start in epididymis then pass along the sperm tube into the urethra and out of the body
4 semen is sperm and fluids to keep sperm alive; sperm fertilise eggs
5 sperm need to be produced at a temperature below normal body temperature
6 if one is damaged or does not work a man can still produce sperm

Unit 5 THE FEMALE REPRODUCTIVE SYSTEM

1 produce eggs, produce female hormones, receive the male sperm, nourish and protect developing baby
2 protect them from damage
3 the external sex organs
4 if one is damaged or does not work a woman can still produce eggs
5 so they can stretch when a woman is pregnant and giving birth
6 only takes one egg to make a baby, sperm have to swim to 'find' the egg

Unit 6 THE MENSTRUAL CYCLE

1 uterus lining is shed
2 a) 5 days; b) the uterus lining breaks down and is passed out of body; c) the uterus wall is contracting and relaxing
3 a) starts an egg maturing; b) follicle bursts and releases egg; c) uterus wall starts to thicken; d) uterus wall thickens more
4 a) egg is released from ovary; b) days 13–15
5 ability to have children

Unit 7 A NEW LIFE BEGINS

1 sperm and egg join/fuse together; occurs in oviduct
2 a) because some die and some cannot swim and it is a long way to the egg (for a sperm cell); b) move their tails back and forth; c) they die
3 24 hours
4 the embryo sinks into the lining of the uterus
5 a) individual e.g. arms and legs grow, nose, mouth and eyes form; b) individual e.g. gets bigger, fingernails grow; c) individual e.g. bones harden
6 fraternal twins come from 2 different eggs fertilised by 2 different sperm; identical twins come from 1 egg fertilised by 1 sperm

Unit 8 BIRTH

1 a) exchanges food and waste between mother and baby; b) carries babies blood to and from placenta; c) protects developing baby
2 individual e.g. drugs and alcohol
3 a) labour – uterus walls contract and relax pushing baby down; b) delivery – baby passes through vagina into the world; c) afterbirth – the placenta is pushed out of the mother's uterus
4 4, 7, 5, 1, 6, 2, 8, 3

Unit 9 REPRODUCTIVE TECHNOLOGY

1 not being able to have children
2 individual e.g. not producing enough sperm, blocked oviducts
3 individual e.g. when sperm and egg fuse in a dish in a lab, not inside a woman
4 she would end up having several babies
5 individual – personal view
6 individual – personal view

Unit 10 SEXUALLY TRANSMITTED INFECTIONS

1 a) sexually transmitted infection; b) venereal disease
2 contact with infected body fluids; unsafe sex
3 people who have many sexual partners, who are sexually active at an early age and people who have sex with an infected person

4 Chlamydia: cause – bacteria, symptoms – e.g. discharge from penis, time to appear – 21 days, treatment – antibiotics
Genital herpes: cause – virus, symptoms – e.g. painful, blistery sores, time to appear – 2–20 days, treatment – none, but antiviral drugs may help
Pubic lice: cause – small insects/lice, symptoms – e.g. itchy anus, time to appear – 5 days, treatment – medication from chemist
HIV: cause – virus, symptoms – e.g. hard to fight infections, time to appear – up to 10 years, treatment – medicines
5 individual e.g. use condoms, abstain from sex

TOPIC 2 ACIDS AND BASES

Unit 11 THE ACID FAMILY

1 a) they have properties in common; b) individual e.g. taste sour, corrosive; c) individual e.g. react with carbonates; d) they are harmful
2 a) individual e.g. hydrochloric HCl; b) lactic, tartaric
3 a) hydrogen; b) hydrogen ions; c) only releases a few hydrogen ions in water

Unit 12 THE BASE FAMILY

1 a) they have properties in common; b) individual e.g. feel slippery; c) individual e.g. cancel out acids; d) an alkali is a base that dissolves in water
2 acid + base ⟶ salt + water
3 individual diagram
4 a) burns: corrosive; b) cleaner and fumes are harmful; c) can produce a poisonous gas

Unit 13 CHEMICAL INDICATORS

1 a) show if something is an acid or a base; b) dyes from plants; c) litmus paper, universal indicator
2 a) the acidity of a solution; b) individual diagram; c) individual diagram
3 a) tea is an acid/base indicator and lemons contain citric acid; b) the soil is pH neutral

Unit 14 ACID-CARBONATE REACTIONS

1 a) carbon dioxide; b) bubble gas through limewater which will then turn cloudy/milky; c) it releases carbon dioxide gas
2 a) a salt + water + carbon dioxide; b) (i) calcium chloride + water + carbon dioxide, (ii) magnesium sulfate + water + carbon dioxide
3 it reacts with stomach acid to produce carbon dioxide so people burp

Unit 15 ACID-BASE REACTIONS

1 a) chemicals that neutralise stomach acid; b) hydrochloric acid HCl
2 a) salt + water; b) (i) sodium chloride + water, (ii) magnesium sulfate + water
3 individual e.g. they pass through the body and out in wastes
4 individual rhyme
5 a) wash it off with plenty of water; b) put on an alkali to neutralise it because they are also corrosive
6 antacids, bee, neutral, water, hydrochloric, wasp, salt, acid, base, vinegar; the 'down' word is neutralise.

Unit 16 ACIDS AND METALS

1 a) a gas is being produced; b) collect gas by holding thumb over end of test tube, light a match and bring it near test tube, remove thumb, listen for pop; c) magnesium + hydrochloric acid ⟶ magnesium chloride + hydrogen
2 acid rain is weaker than lab acid
3 a) lead + sulfuric acid ⟶ lead sulfate + hydrogen; b) prevents hydrogen explosions, the lead sulfate dissolves in it
4 individual diagram

Unit 17 WHAT IS 'SALT'?

1 individual e.g. something made when an acid and base react
2 hydrochloric, sulfuric, nitric
3 a) (i) hydrochloric acid, magnesium, (ii) sulfuric acid, calcium
b) Mg + HCl ⟶ MgCl + H (not balanced)
4 individual diagram
5 individual e.g. on food, in fertilisers

Unit 18 ACIDS, BASES AND THE HUMAN BODY

1 a) bacteria; b) makes you produce more saliva that neutralises plaque acid
2 a) lactic acid builds up in muscles; b) stop exercising; c) producing energy using oxygen
3 a) strong, it has pH of 2; b) kill bacteria, help the enzyme pepsin work; c) neutralise stomach acid
4 bases/alkalis
5 neutralises acidity of skin allowing bacteria to grow
6 a) help it work properly; b) pepsin, breaks down protein
7 drink water

Unit 19 ACIDS, BASES AND pH IN OUR EVERYDAY LIVES

1 carbonic or phosphoric acids, calcium carbonate, sodium hydroxide, sulfuric acid
2 a) H_2SO_4; b) onions, fertiliser, car battery
3 a) base; b) attaches to dirt to remove it from skin, neutralises skin acidity
4 a) superphosphate; b) lime

TOPIC 3 ELECTRICITY

Unit 20 WHAT IS ELECTRICITY?

1 individual e.g. light, TV
2 a) non-moving electric charge; b) friction on a surface adding or removing electrons
3 a) electrons; b) negative
4 individual e.g. pushes against
5 a) individual – any metal object; b) metals
6 a) individual – non metals; b) individual e.g. no, some are wood and some are plastic

Unit 21 HOW IS ELECTRICITY MADE?

1 a) hydroelectric power stations – water; b) wind; c) they are renewable
2 moving water, mechanical, electrical
3 a) 500 kV; b) 25 kV; c) 0.5 kV
4 a) boosts or increases voltage; b) drops or decreases voltage
5 a) 110 kV , 110 kV, 33 kV, 11 kV, 240 volts, 240 volts; b) because the current has to travel long distances
6 a) chemical reaction; b) can be recharged

Unit 22 ELECTRIC CIRCUITS

1 a) supply of electrical energy, unbroken pathway, components to transform energy; b) component (resistor)
2 a) easier and quicker; b) power pack, lamp, ammeter, switch
3 a) individual diagram; b) individual diagram
4 individual, e.g. series has only one pathway for current and parallel has more than one pathway

Unit 23 ELECTRIC CURRENT

1 a) individual e.g. electrons moving in a wire; b) conventional current moves from positive to negative and actual current negative to positive; c) symbol = I unit = amps, instrument = ammeter, electrons = 6×10^{18}, DC = direct current
2 a) positive of ammeter connected to negative of power pack, b) ammeter is parallel not series
3 5.5 amps, 8 amps, 2.75 amps
4 a) all lamps go out; b) another pathway for current to move along
5 can cause a fire or electrocution

Unit 24 VOLTAGE

1 a) energy; b) power supply e.g. power pack
2 voltage = energy, symbol = V, unit = volts, instrument = voltmeter, joule = unit of energy
3 a) to measure energy used by a component; b) to stop most of current going through it; c) current would go through it instead of main circuit
4 a) voltmeter in series, b) positive of voltmeter connected to negative of power pack
5 a) parallel circuit – because they do not have to share the energy; b) yes – they will get dimmer

Unit 25 ELECTRIC POWER

1 individual e.g. how much energy something in a circuit changes
2 a) how quickly it changes electrical energy into microwave energy; b) uses less electricity so pay less
3 a) 200 x 3600 = 720,000 joules; b) 0.2 kW x 1 hour = 0.2 units x 20c = 4 cents
4 993 (top right of bill)

Unit 26 ELECTRICITY AND THE HUMAN BODY

1 human nervous system uses small electric currents
2 a) individual e.g. burn, cause heart attack; b) start their heart beating when it has stopped; pacemaker
3 big shock with wet hands, smaller shock with dry hands
4 a) so someone else can call for ambulance/medical help; b) airway, breathing, circulation

Unit 27 SAFETY WITH ELECTRICITY

1 break circuit so current stops flowing
2 needs less current/voltage to run lights
3 a) live, neutral, earth; b) live – it brings current into appliance; c) safety – takes current to earth
4 a) when using electrical equipment outside; b) protect you from electric shock
5 individual answers e.g. do not touch plugs/switches with wet hands

TOPIC 4 EARTH'S CHANGING FACE

Unit 28 EARTH'S STRUCTURE AND CONTINENTAL DRIFT

1 studying seismic waves, studying rocks
2 a) under the land; b) the mantle; c) high pressure due to other layers above it
3 a) Pangaea; b) Laurasia; c) Gondwanaland
4 a) individual e.g. why all the continents are where they are; b) individual e.g. fossils
5 a) a person who studies the Earth's crust (rocks); b) a person who studies the atmosphere and weather; c) a person who studies living things

Unit 29 PLATE TECTONICS

1 a) pieces of crust with continents embedded in them; b) coming together; c) moving apart; d) moving underneath something else
2 a) Pacific and Indo-Australian; b) individual e.g. Antarctic, Eurasian; c) Nazca
3 a) faults develop and magma comes out e.g. Atlantic Ocean floor; b) trenches form e.g. Mariana trench; c) earthquakes occur e.g. name of any earthquake
4 The Ring of Fire occurs where many plates meet

Unit 30 EARTHQUAKES

1 a) individual e.g. a vibration travelling through the ground; b) cracks and fractures in Earth's crust
2 a) tectonic plates moving, rocks pushed together under pressure move; b) tsunamis, avalanches, landslides; c) move up or down or sideways
3 a) individual general knowledge; b) individual general knowledge
4 individual diagram
5 a) P waves are faster than S waves; P waves push and pull the ground and S waves make the ground roll; b) P and S waves move through the Earth and the others move on the surface
6 a) size of ground movement; b) effect on people and buildings; c) total energy released

Unit 31 VOLCANOES

1 a) an opening in Earth's crust; b) because these are weak points in the crust; c) how easily the lava flows
2 individual diagram

3 dome e.g. Mt Tarawera, shield e.g. Rangitoto, cone e.g. Mt Taranaki
4 individual e.g. Ruapehu, Ngauruhoe, Tongariro, Pihanga

Unit 32 WEATHERING AND EROSION

1 a) individual e.g. breaking down of rock; b) individual, e.g. carrying pieces of rock away
2 chemical, biological, mechanical
3 a) frozen ice gouges rub away rock on valley walls and floor; b) edges get smoothed as rock bumps along in river; c) roots grow and break it up; d) roots hold the soil and stop it being washed or blown away
4 individual map

Unit 33 NEW ZEALAND'S GEOLOGICAL HISTORY

1 a) over 4 billion years old; b) around 500 million years old
2 a) 85 million years ago; b) 2.4 million years ago; c) 35 million years ago
3 individual diagram

4U2DO

1 Acid

a) What produces the acid that causes tooth decay? ______

b) One chewing gum maker says chewing their sugar-free gum helps fight plaque acid. Why would it work?

______

2 Exercise

a) Why do people suffer from stitch when they exercise?

______

b) What do they have to do to make the stitch go away?

______

c) What does aerobic respiration mean?

______

3 Stomach acid

a) Is stomach acid a strong or weak acid? How do you know?

______

b) Give two reasons why the stomach produces acid.

______

c) How does taking antacids soothe the pain caused by stomach ulcers?

______

4 Bicarbonate ions in the blood help reduce the acidity of blood. What family of chemicals must bicarbonate belong to?

______

5 Explain how the use of an alkaline soap may increase the chance of getting an infection through the pores in your skin?

______

______

6 Enzymes

a) What do enzymes do in the body?

______

b) Name one enzyme mentioned in this unit and say what it does.

______

7 What do people do during the day that they do not do during the night that may affect the pH of urine?

______

8 Write down three things you have learned from this unit.

a) ______

b) ______

c) ______

9 Write down anything you need to ask your teacher to explain.

______

UNIT 19 Acids, Bases and pH in our Everyday Lives

Carbonic acid, H_2CO_3 Phosphoric acid, H_3PO_4 	To make fizzy drinks fizz, carbon dioxide is added to water. In water, carbon dioxide forms carbonic acid. Phosphoric acid is sometimes added to soft drinks because it helps the drink absorb carbon dioxide. Hydrogen ions from the acids give the drink its sharp taste and a pH of about 4.
Acetic acid (vinegar), CH_3COOH 	Vinegar has been used for hundreds of years to preserve food in a method called pickling. The acetic acid stops bacteria and fungi growing on the food and digesting it for their own energy needs.
Hypochlorous acid, HClO 	Chlorine is added to drinking water to kill disease-causing bacteria. Chlorine is soluble in water and forms a mixture of hydrochloric and hypochlorous acids. Hypochlorous acid is a weak inorganic acid that breaks down when exposed to sunlight and heat.
Baking soda, $NaHCO_3$ 	Baking soda in toothpaste helps neutralise the acid produced by bacteria on your teeth. It also acts as a polishing agent because it is coarse and abrasive.
Sodium hypochlorite, NaClO 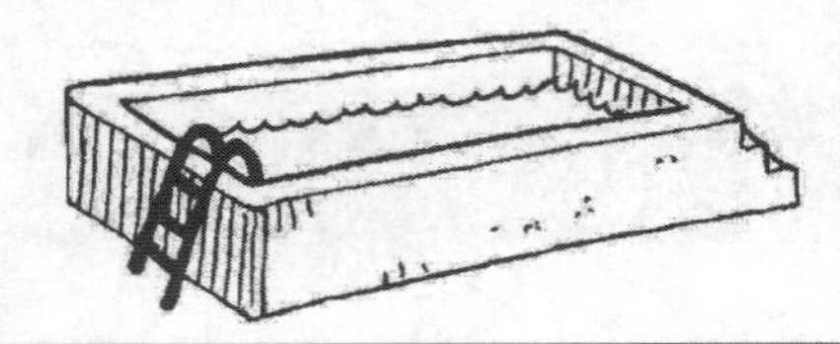	Keeping the pH of your swimming pool at the right level is important for several reasons: to keep control of bacterial and algal growth, to prevent the pool water corroding the pool lining, and to stop eyes stinging when you swim.
Sulfuric acid, H_2SO_4 	When you peel or chop onions, enzymes in the damaged onion cells react with other substances in the onion to release a gas that dissolves in the water of your eyes to form sulfuric acid. The acid causes your eyes to sting. Then your brain makes you cry to dilute the acid but this means there is more water for the gas to dissolve in so more acid is made.
Sodium hydroxide, NaOH 	Soap is made from sodium hydroxide and a fat or oil. When soap dissolves in water, the surfactant (surface active agent) attaches to dirt and pulls it from the skin. The dirt is then removed in the dirty water.
Limestone, $CaCO_3$ 	Natural soil is slightly acidic. Many ions that plants need for healthy growth are only soluble when the soil is slightly acidic. Fertilisers such as superphosphate are acidic and add to the soil's acidity; so does rain. If soil becomes too acidic, gardeners apply lime (crushed limestone, $CaCO_3$) to the soil. Different plants like different soil acidities, for example tomatoes prefer a pH of 6–7, beans pH 5.5–7, cabbages and roses pH 6–6.5.
Sulfuric acid, H_2SO_4 	Plants and animals need phosphorus for normal growth. For people it is an important part of bones and teeth. Plants get it from the soil and people get it from eating plants or eating meat and milk. Rock phosphate is insoluble until it has been treated with sulfuric acid. This reaction produces soluble superphosphate, $Ca(H_2PO_4)_2$, which is added to soil as a fertiliser.

Sulfuric acid, H_2SO_4 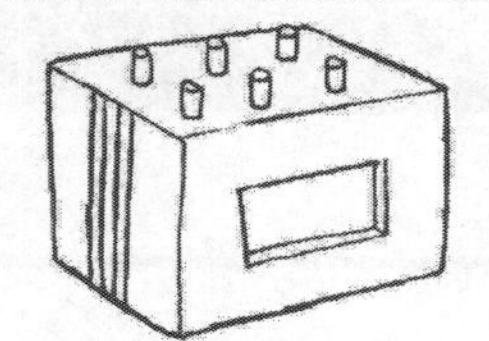	Car batteries are lead-acid batteries. They are made up of six two-volt cells. Each cell is made of two lead electrodes separated by plastic sheets. Sulfuric acid surrounds the electrodes. The acid reacts with the electrodes to produce the electric current necessary to start the engine.
Sulfurous acid, H_2SO_3 	Acid rain is more acidic than normal rain because acidic gases such as sulfur dioxide (produced when fossil fuels are burned) dissolve in water droplets to form sulfurous acid.

4U2DO

1 For each of the diagrams below, write the name of the acid or base it contains.

			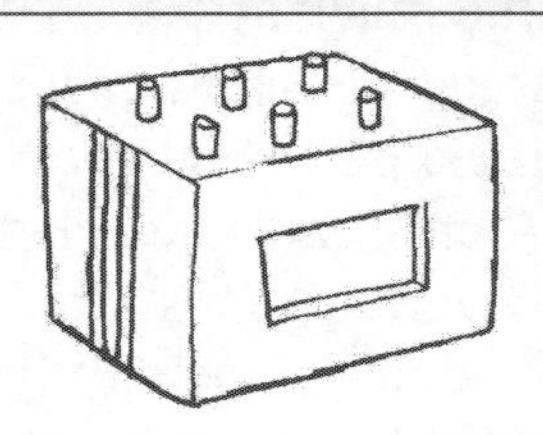

2 Sulfuric acid is linked to many different products.

a) Write the chemical formula for sulfuric acid.

____

b) Name three different products it is linked to.

____

3 Soap

a) Is the sodium hydroxide in soap an acid or a base?

____

b) Describe two ways soap affects your skin.

____

____

4 Soil

a) A gardener wants to grow plants that prefer an acidic soil. What should be added to the garden before planting takes place?

____

b) In another garden, hydrangeas with blue petals are growing. To make the petals red the gardener needs to make the soil neutral. What should be added to the soil?

____

5 Write down three things you have learned from this unit.

a) ______________________________

b) ______________________________

c) ______________________________

6 Write down anything you need to ask your teacher to explain.

____

UNIT 20 WHAT IS ELECTRICITY?

What do you think of when you hear the word 'electricity'? Lights, TV, and computer games? When we think of electricity we tend to think of devices that use electricity but electricity is much more than this. Electricity is a form of energy. It provides the power to do many kinds of work.

Electricity is a basic feature of atoms. Atoms are the building blocks of matter.

The nucleus of an atom contains positively charged protons. Negatively charged electrons move around the nucleus. These two oppositely electrically charged particles attract one another and this holds the atom together.

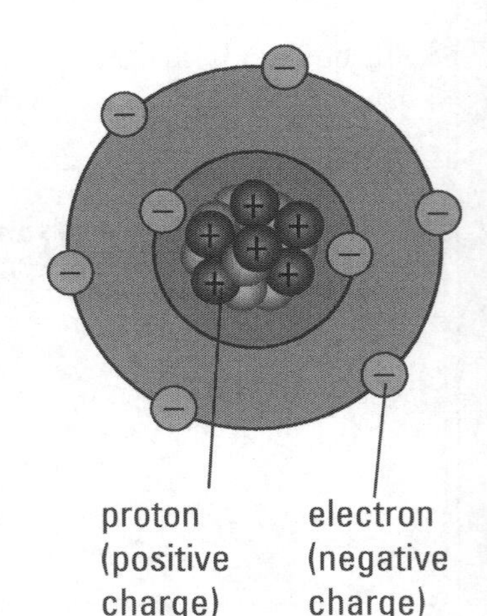

Atoms can gain electrons to form negative ions or lose electrons to form positive ions. Ions that have opposite electrical charges attract one another to form ionic compounds. Table salt is an ionic compound formed when a sodium ion (positive) and a chloride ion (negative) join together.

An object has a static (non-moving) electric charge when a large number of the atoms that make up the object have gained or lost electrons. Examples of static electricity include:

- the amber that the early Greeks discovered which attracted bits of feathers and dust
- when you rub a balloon on your hair then hold it near your head and it makes your hair stand on end
- when nylon clothes crackle as you pull them apart after they have been in the clothes dryer
- when your hair stands on end as you jump on a trampoline
- lightning that jumps between a positively charged cloud and a negatively charged cloud, or a cloud and the ground.

A power supply such as a battery gives electrons energy to move from one atom to another atom. Electrons 'kicked out' of one atom move to the space in another atom that has been left by an electron that has been 'kicked out' from there. The charge is passed from atom to atom like passing a ball along a chain of people. This movement of electrons is an electric current.

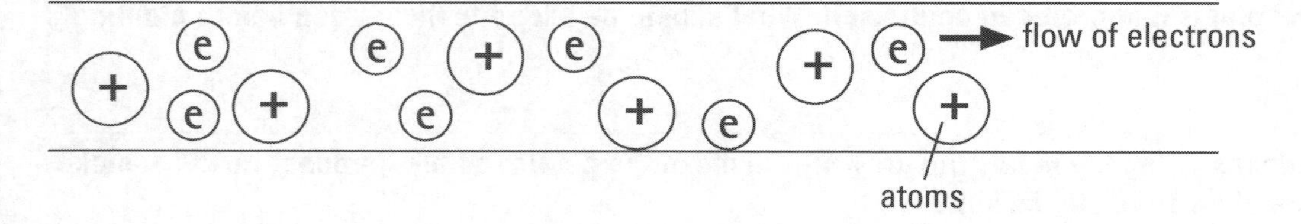

The term resistance refers to how difficult it is for electrons to move through a material or component.

Materials are called conductors if they let electrons move through them. Scientists say they have low resistance. Metals are good conductors. This is because they contain electrons that are not strongly attached to an atom so they are free to move. When a battery is connected to a metal, it provides the energy that gives these free electrons a 'push' and they start to move. An electric current is made up of moving electrons (see Unit 22).

As electrons pass through a conducting wire, some of their energy makes the atoms in the wire vibrate. This makes the wire get hot. The more resistance a wire has, the hotter it will get. In a light bulb, the filament you can see is a 'coiled

> Amber is fossilised tree sap that is gold in colour. The early Greeks discovered that when amber was rubbed, it attracted bits of lightweight things like feathers and dust. They called amber *elektron* (from the word *elektor*, meaning 'beaming sun').
>
> The Latin word *electricus* means to 'produce from amber by friction'. We get the words electron and electricity from these Greek and Latin words.

coil' of tungsten wire. When an electric current passes through the filament, about 10 per cent of the electrical energy is transformed into useful light. The rest is transformed into heat.

Materials are called insulators if they do not let electrons move through them. Insulators have a high resistance (they oppose the flow of electrons). The atoms of insulators hold on tight to their electrons. Their electrons stay in place even when they are connected to a battery. Plastic, wood and rubber are all good insulators.

1 Name as many things around you right now that use electricity.

2 Static electricity

a) What does static electricity mean?

b) How do objects become statically charged?

3 Atom part

a) What part of an atom moves to form an electric current?

b) What charge does this part of the atom have?

4 In your own words, explain what resistance means.

5 Conductors

a) Name three electrical conductors in your classroom.

b) What physical feature do all these conductors have in common?

6 Insulators

a) Name three electrical insulators in your classroom.

b) Are these insulators all made of the same material? If not, what are they made of?

7 Write down three things you have learned from this unit.

a)

b)

c)

8 Write down anything you need to ask your teacher to explain.

UNIT 21 How is Electricity Made?

Wind turbines

Electricity is one of our most commonly used energy sources. Electricity is generated in a power station. In New Zealand:

- about 10 per cent of coal mined is used in power stations such as the Huntly power station to produce steam to drive the turbines.
- over 40 per cent of natural gas is used for the same purpose. The Huntly power station can burn natural gas as well as coal.
- 70–80 per cent of electricity demand is produced from water power in hydroelectric stations.
- about 5 per cent of electricity demand is met by geothermal generation.
- wind power has the potential to supply 10 per cent of electricity needs.

In any electric power station, the basic process of creating electricity is the same. A turbine, which is a wheel or cylinder with blades arranged around its edge, converts the energy of steam or moving water into mechanical energy.

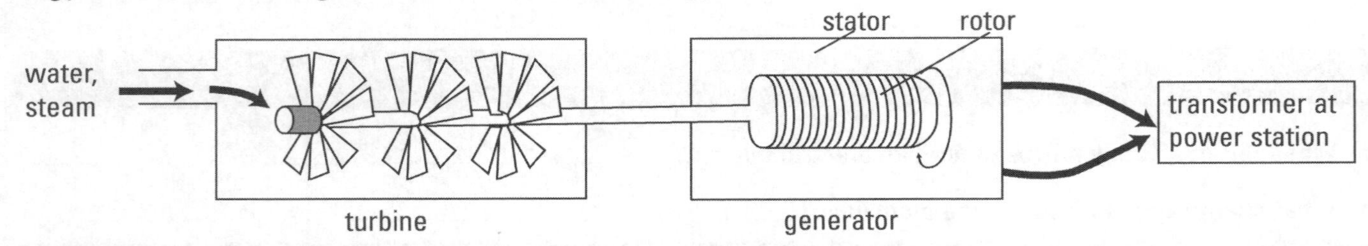

A generator connected to the turbine then converts the mechanical energy into electrical energy. The generator uses a rotor (the moving part), which is a series of wire coils with a small electric current running through them. This current produces a magnetic field around the coils. These coils spin (because of the turbines) inside the fixed part of the generator, the stator, which is also a series of coils. The spinning magnetic field of the rotor induces a large electric current to flow in the stator coils. The electricity made in the stator has a voltage of about 11,000 volts (11 kilovolts, or 11kV). (Voltage is a measure of the electric force that 'pushes' electrons around a circuit. See Unit 23.)

The diagram shows what happens to the electricity before it gets to your television.

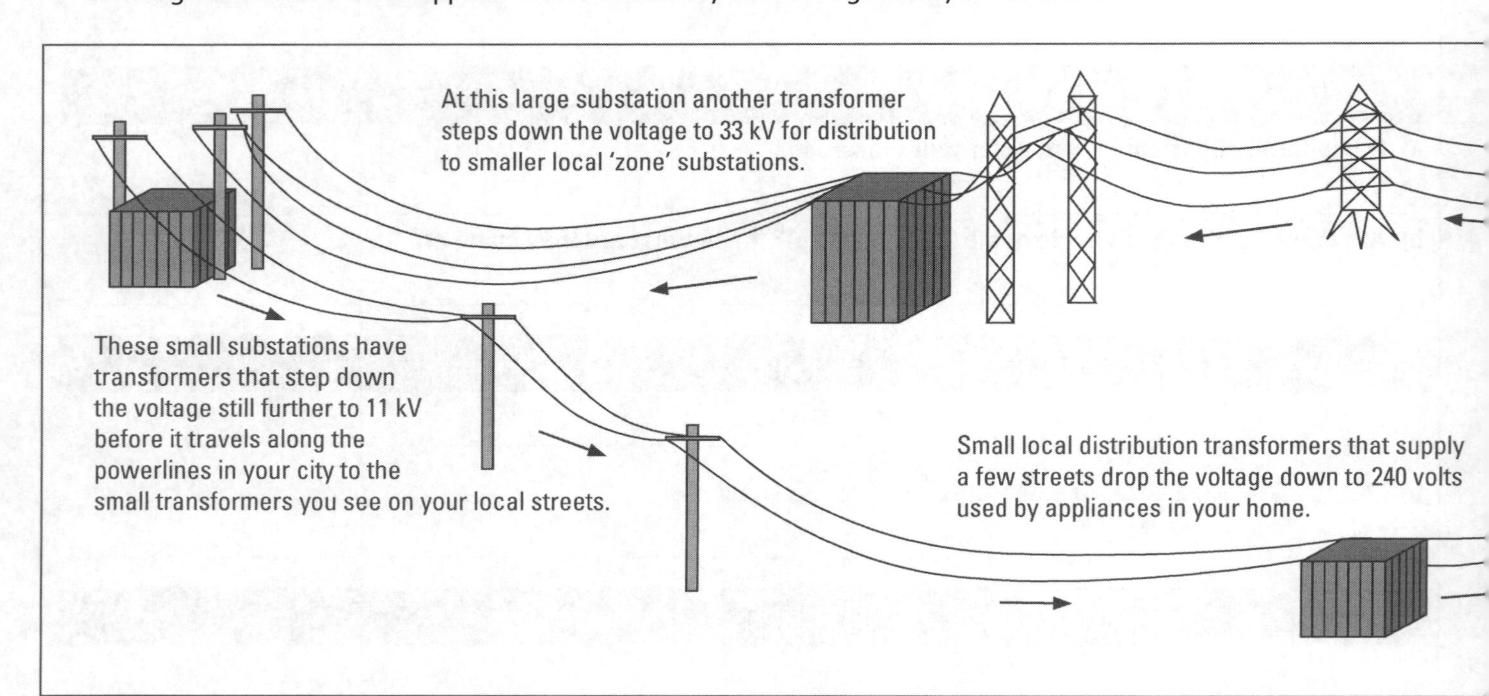

Many things you use around your home do not get their electrical energy from the wall socket but from batteries. Batteries transform chemical energy into electrical energy. The basic unit of a battery is an electrochemical cell. The diagram shows a simple, inexpensive type of battery commonly used in torches.

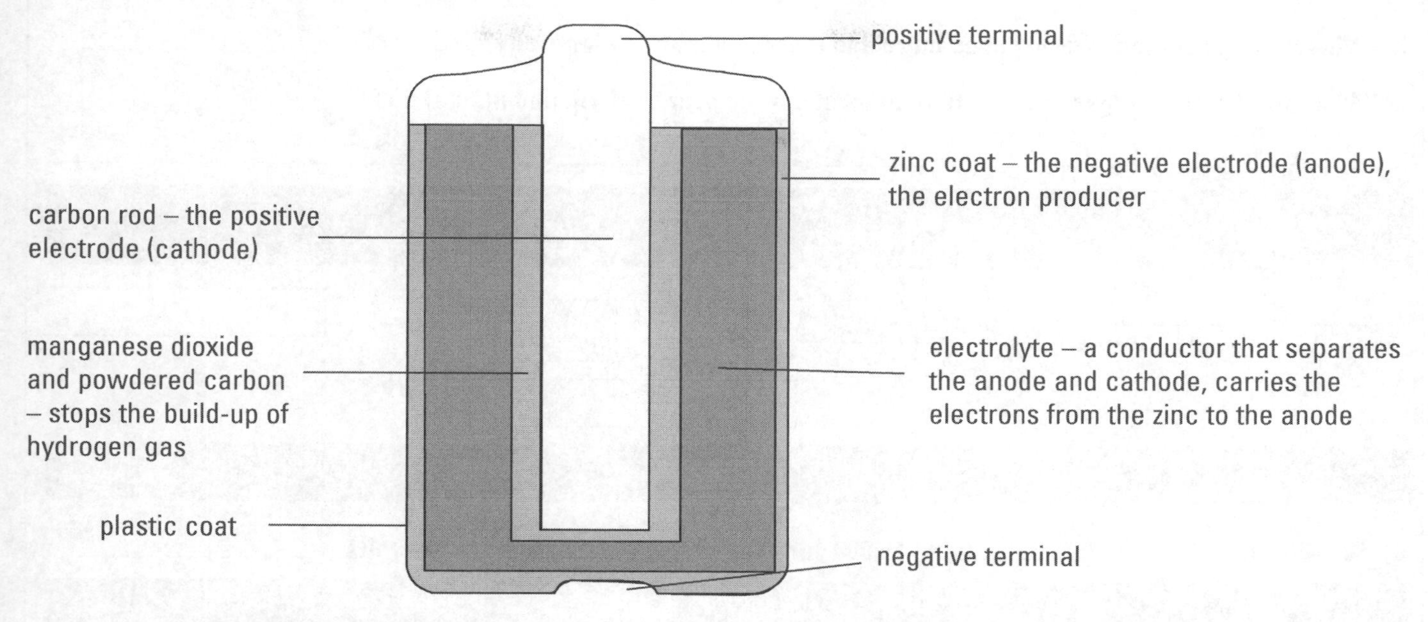

In this type of battery, electrons only flow when the positive and negative terminals are connected to complete a circuit. This is what happens when you turn on the switch.

There are several types of batteries:

- alkaline batteries (Duracell and Energizer) use an alkaline electrolyte such as potassium hydroxide. They are used in high-energy toys and games.
- lithium batteries use lithium metal as their anode. They produce twice the voltage of an alkaline battery of the same size. They are used in cameras and watches.
- nickel-cadmium or NiCad batteries have nickel hydroxide and cadmium electrodes. They can be recharged (this reverses the chemical reactions, returning the battery to its original condition). They are used in power drills.

Batteries come in different sizes, D (largest), C, AA and AAA (smallest). They may produce varying voltages (electrical energy). They may be made up of one (electrochemical) cell or many cells.

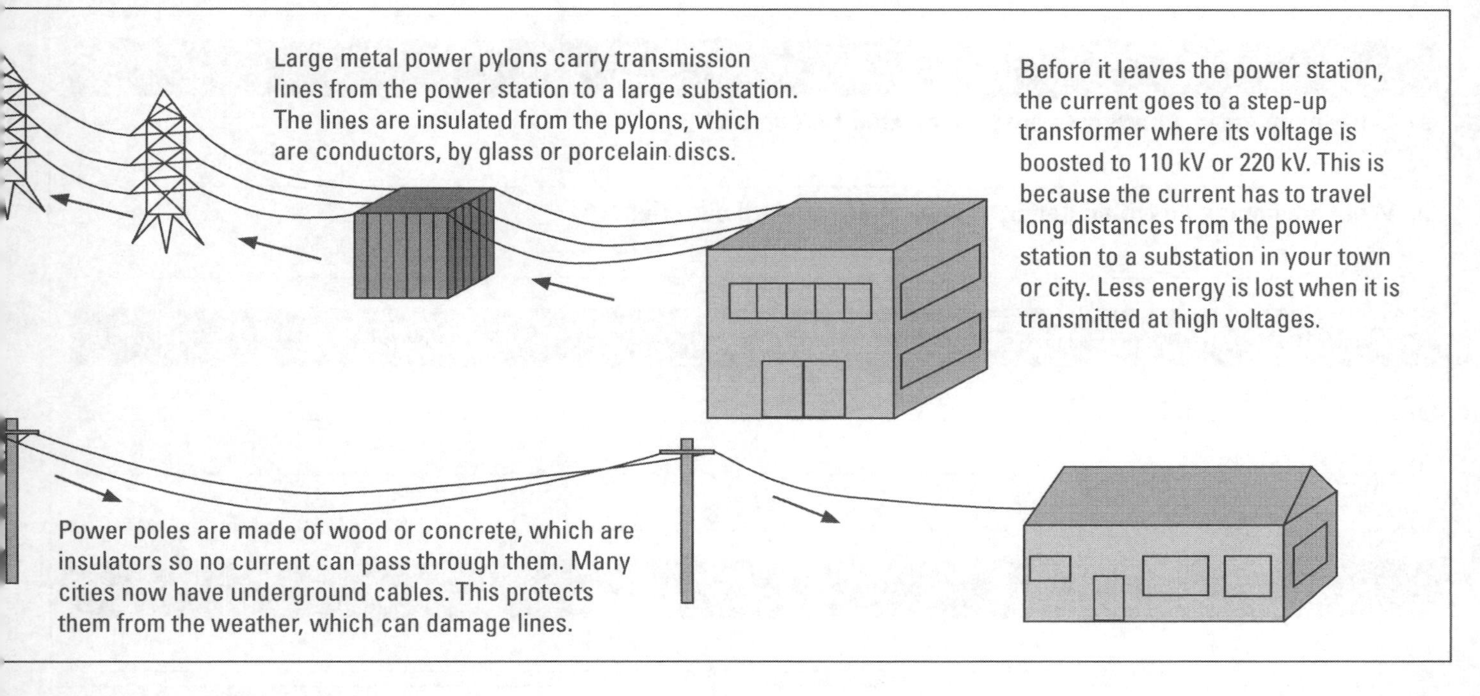

4U2DO

1 Forms of energy

a) In New Zealand, which form of energy is used to generate most of the electricity we use? ______

b) Which energy source is being used more and more to generate electricity? ______

c) What are the advantages of these two forms of energy over coal, oil and natural gas?

______

2 Fill in the boxes to show the three different energy transformations that take place in a hydroelectric power station.

turbine

water

generator

3 A kilovolt is 1000 volts. Convert these voltages to kV.

a) 500,000 volts ______ b) 25,000 volts ______ c) 500 volts ______

4 In your own words explain

a) what a step-up transformer does ______

b) what a step-down transformer does. ______

5 Voltage

a) Complete this flow diagram by writing in the voltage at each point.

produced by generator	in transmission lines	large substation	small substation	street transformer	your house

b) Why is the voltage so high in transmission lines?

______

6 Batteries

a) What happens in a battery to produce an electric current?

______

b) What advantage do NiCad batteries have over standard batteries?

______

7 Write down three things you have learned from this unit.

a) ______

b) ______

c) ______

8 Write down anything you need to ask your teacher to explain.

______

UNIT 22 ELECTRIC CIRCUITS

A circuit is an unbroken pathway that electrons (electric current) can flow through. If there is a gap in the circuit, electrons will not flow.

There are three basic parts to an electrical circuit:

1. a supply of electrical energy for the moving electrons (current), for example a single cell, a battery (two or more cells together), or a power pack
2. an unbroken pathway for the electrons to flow through, for example wires
3. components to transform the electrical energy into other forms of energy, for example lamp, bell, motor.

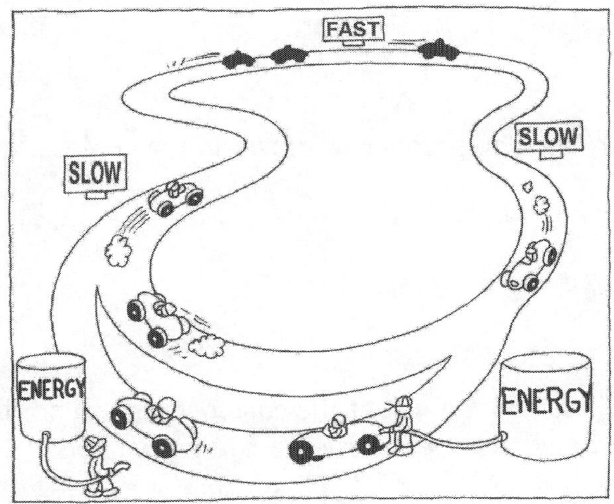

Imagine an electric circuit is like a car racetrack. The pit area is the power supply. Here the crew fill the car (the electrons) with petrol (the energy). The track has fast straights and slow corners. It may also have hills. Cars (electrons) use up their fuel (energy) as they race around the track.

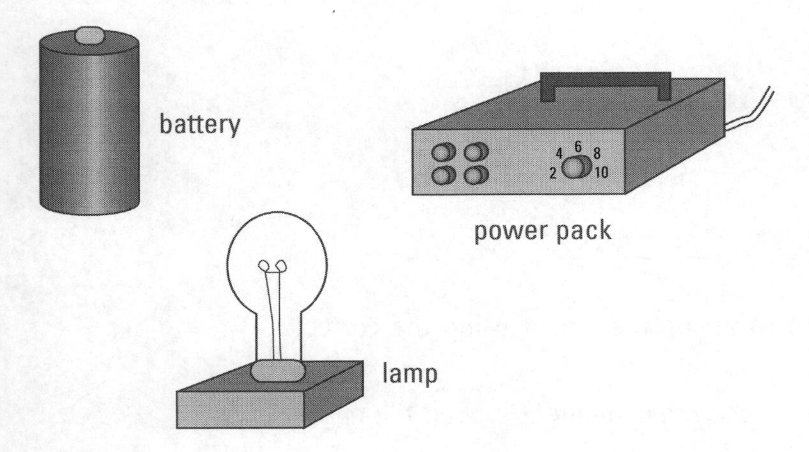

Other things that can be put into a circuit include:

- measuring devices, for example ammeter and voltmeter
- current controllers, for example switches, fuses.

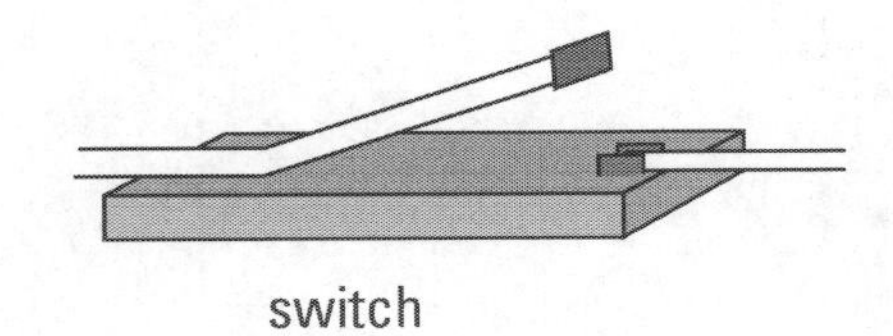

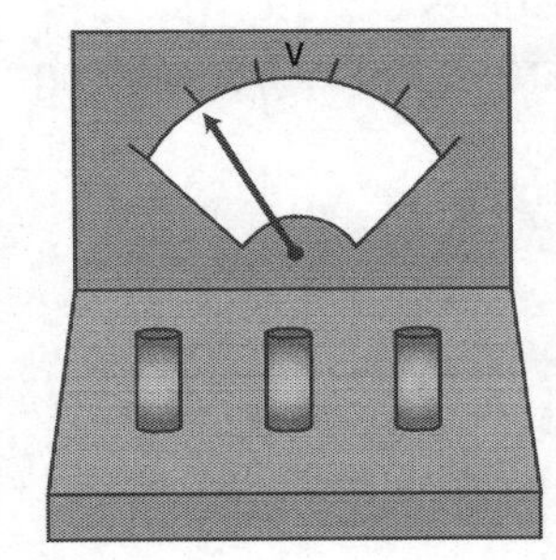

It would be a long and difficult task to draw all the equipment used in an electric circuit so scientists use symbols to represent each different component.

Component	Symbol	Component	Symbol	Component	Symbol
cell	+ −	wire		ammeter	A
battery	+ −	open switch		voltmeter	V
power pack		lamp		fuse	

HOW TO DRAW A CIRCUIT

1 Circuits are drawn in the shape of a rectangle. Use a ruler to draw the lines that represent the connecting wires.

2 The power supply is drawn in the centre of the top side of the rectangle.

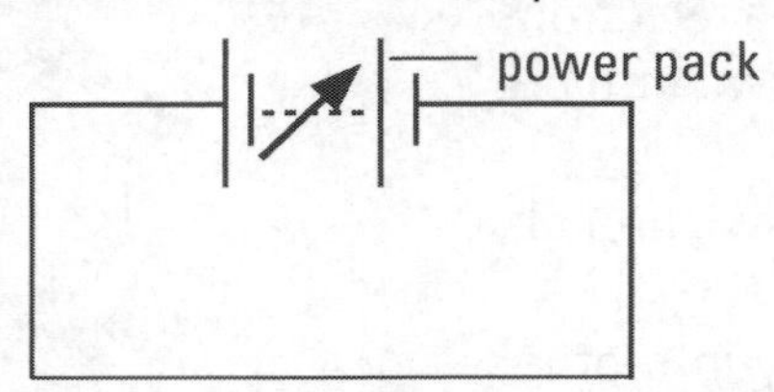

3 Spread the rest of the components around the other sides of the rectangle. Do not put any at a corner. Use the correct symbols.

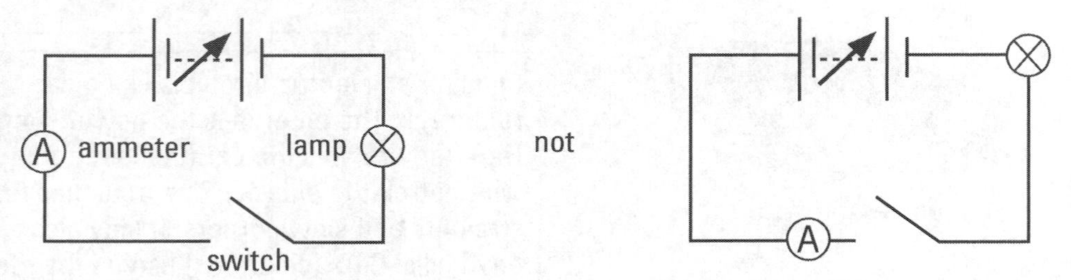

When you are given a circuit diagram and asked to set up the circuit using the correct equipment:

- first picture a rectangle on your desk and then place the equipment around the edges in the positions shown in the circuit diagram
- next connect each component with wires so an unbroken pathway is made.

Series and parallel circuits

In a series circuit, the lamps are connected one after the other and they all light up as the electrons move through them.

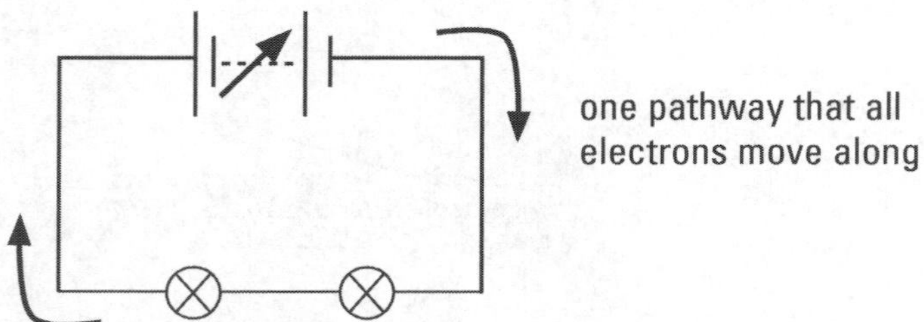

In a parallel circuit the lamps are connected in parallel pathways to each other so there is more than one pathway or circuit. The lamps all light up but their brightness is different to when they are connected in series. (See the next two units to find out why.)

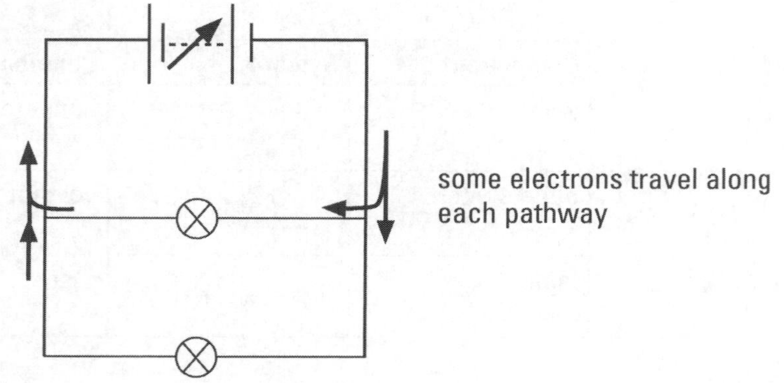

4U2DO

1 Circuit

a) List the three basic requirements for an electric circuit.

b) In the car comparison, which part of the circuit is represented by the hill?

2 Symbols

a) Why are circuit symbols used when drawing circuit diagrams?

b) Name these circuit symbols.

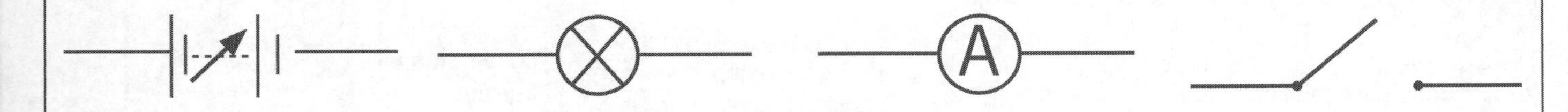

3 Draw a circuit diagram to show the following:

a) a power pack connected to a lamp in series with a switch and an ammeter

b) a battery connected to two lamps in parallel with a switch to turn off both lights at the same time.

4 In your own words, explain the main difference between a series and a parallel circuit.

5 Write down three things you have learned from this unit.

a)

b)

c)

6 Write down anything you need to ask your teacher to explain.

UNIT 23 Electric Current

An electric current is a flow of electrons. The negative terminal of a power pack 'pushes' electrons out into the circuit where they move from atom to atom in the wires to the positive terminal of the power pack.

As they move along a wire, the electrons bump into atoms. This makes them vibrate faster than normal, and this causes the wire to heat up. Some things like light bulbs and heating elements use this heat to light up or heat things.

Batteries and power packs produce a current that travels in only one direction. This is called direct current (DC).

Before scientists knew what electrons were and how they moved, they mistakenly believed that an electric current flowed from positive to negative. This became a rule, called *conventional current*, and no one has bothered to change it.

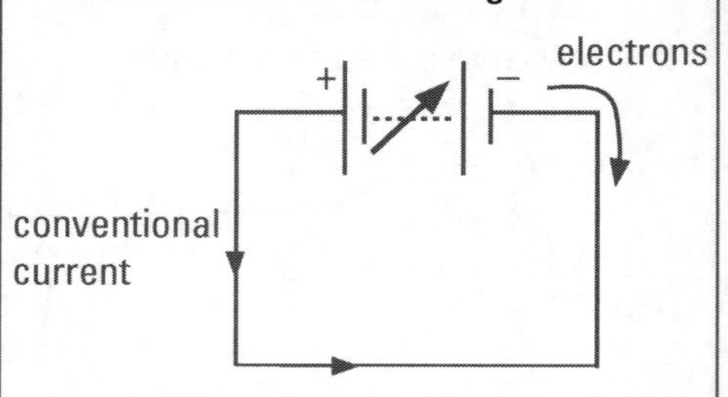

CURRENT

The symbol for electric current is 'I' (the I stands for intensity).

Current is measured in amperes (amps), symbol A.

If an ammeter shows a current of 1 amp, it means 1 coulomb of electrons is passing a point in the circuit every second. So the larger the current, the faster the rate of flow of electrons.

An ammeter 'counts' the number of electrons that pass a point in the circuit every second. Rather than count the electrons individually (because they are so small), the ammeter counts them in groups of 6,241,500,000,000,000,000 (that is, about 6×10^{18}) at a time. This huge number of electrons is called a coulomb.

An ammeter is always connected in series in a circuit. This is so it can measure the current as it passes through it. It has a low internal resistance so the current passes easily through it without losing much energy.

The positive (red) terminal is either connected directly to the positive terminal of the power pack or leads to the positive terminal of the power pack.

The negative (black) terminal is connected so it is leads back to the negative terminal of the power pack.

Some ammeters have two scales. The scale you read depends on which terminals are being used in the circuit. Always start using the largest scale because the internal mechanisms of ammeters are easily broken.

The connecting wires in a circuit are like the racetrack described in the previous unit. Many electrons (cars) move along them at one time. The wider the track the more electrons (cars) that can move around the circuit at one time. Components in the circuit are like corners along the track. The electrons have to slow down and go in single file to get around the corner without crashing. As the electrons slow down, the energy they carry is transformed into other forms of energy.

In a series circuit, there is only one pathway for the electrons to flow through, so all of the current passes through each component.

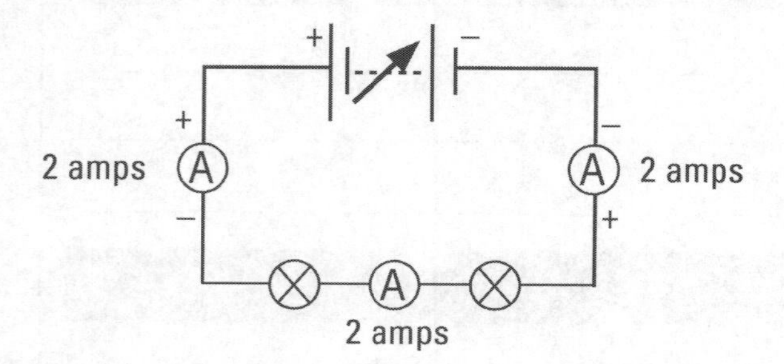

In a parallel circuit, there is more than one pathway for the electrons to flow through, so the current divides and some current passes through each pathway.

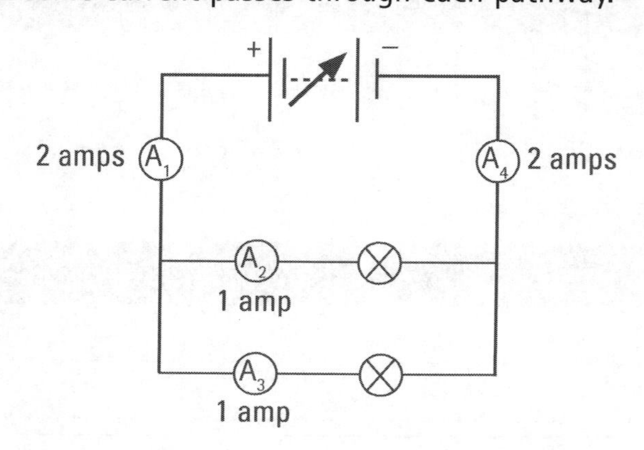

Current will always take the path of least resistance. A short circuit occurs when the current finds a way to bypass the appliance on a path that has little or no resistance, for example where frayed insulation bares a wire and allows it to touch the frame of the appliance so the current can flow straight to ground. In this situation, a very large current can occur, producing a lot of heat and a fire hazard.

4U2DO

1 Current

a) In your own words describe an electric current.

b) Explain the difference between conventional current and 'actual current'.

c) Complete this table about current.

symbol		number of electrons in 1 coulomb	
unit		what DC stands for	
instrument that measures it			

2 Explain what is wrong with these circuit diagrams.

a)

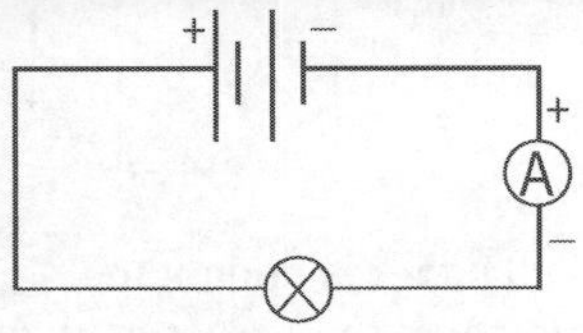

b)

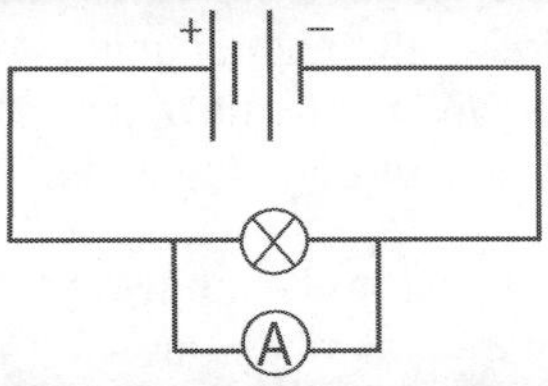

3 What are the readings on these ammeters?

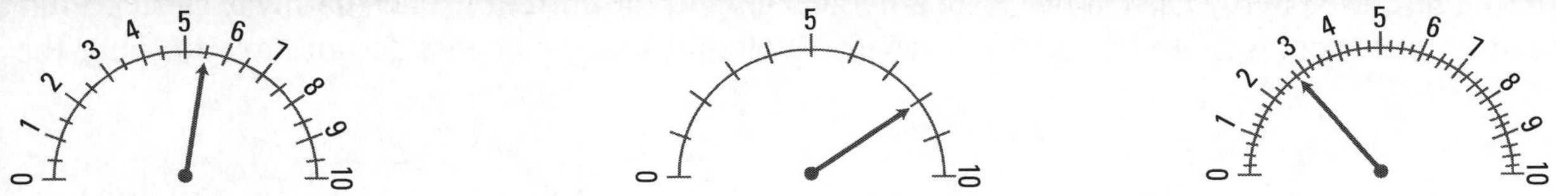

4 Lamps in circuit

a) What happens to the lamps in a series circuit if one of them 'blows'?

b) Why doesn't the same thing happen when the lamps are connected in parallel?

5 Explain why a short circuit can be dangerous.

6 Write down three things you have learned from this unit.

a)

b)

c)

7 Write down anything you need to ask your teacher to explain.

UNIT 24 VOLTAGE

A battery or power pack 'pushes' electrons around a circuit. In a battery the energy to do this comes from the chemical energy of the battery. In a power pack it comes from electrical energy. Electrons gain energy as they pass through the battery/power pack and then 'lose' this energy as the electrons move around the circuit and components like lamps transform it into other forms of energy. It is important to remember that it is the energy the electrons 'carry' that is used by the components, not the electrons themselves. The electrons return to the battery or power pack and pick up more energy before travelling around the circuit again.

The energy gained and lost in a circuit is called voltage. The symbol for voltage is V and it is measured in volts.

In a circuit, voltage is measured using an instrument called a voltmeter.

If a voltmeter is connected to a lamp and it shows a reading of 1 volt, it means each coulomb of current passing through the lamp has given up 1 joule of energy to make the lamp light up (energy is measured in joules).

- In a circuit, a voltmeter measures the energy gained by the electrons as they pass through the power pack or 'lost' as they pass through a component. To do this, a voltmeter must be connected in parallel so it can measure the difference between the energy going into and the energy coming out of a component. Voltmeters have a large internal resistance so very little current passes through the meter itself.

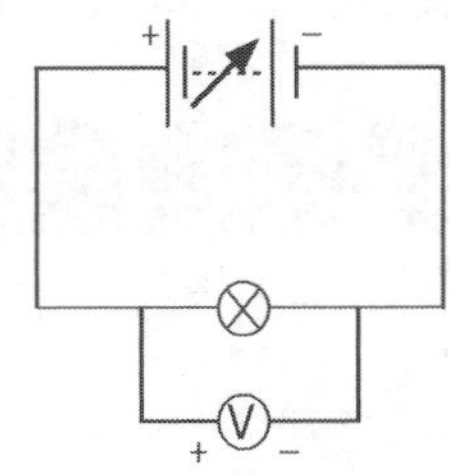

- The positive (red) terminal either is connected directly to the positive terminal of the power pack or it leads to the positive terminal.
- The negative (black) terminal either is connected directly to the negative terminal of the power pack or it leads back to the negative terminal.

In a series circuit, the energy carried by the current (the voltage) is shared by all the components.

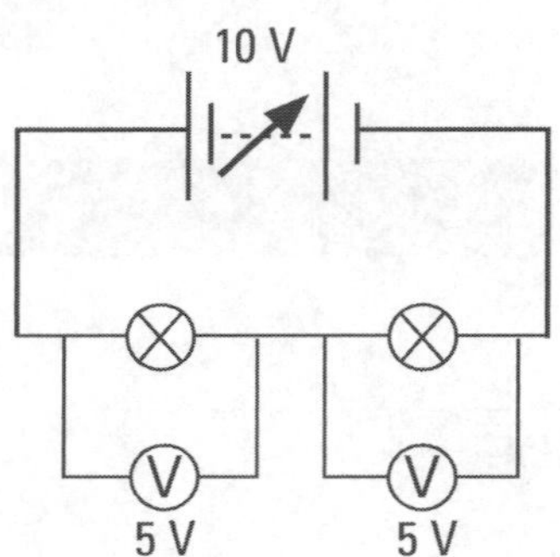

In a parallel circuit, the current in each pathway carries a full amount of energy (voltage) so components do not have to share the energy.

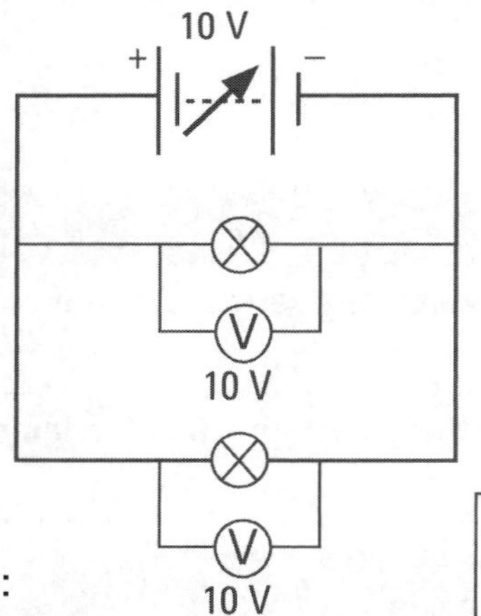

This may be easier to understand if you think of it this way:

Mum (the power pack) gives her two children (the current) $10 each (the energy) to spend at the shopping mall.

When they get to the mall, one child goes up the escalator to the top floor and spends all of the $10 at the video store.

The second child stays on the ground floor and spends $2 at the sweet store and $8 at the bookstore.

You can see that the current (the children) had two pathways to choose from (the two floors of the mall), so some current (one child) went along each pathway. The amount of energy or voltage (the money) in each pathway is the same ($10) and it is used by the components in the pathway (the shops).

The current (children) were not used up. They went home and asked Mum for more spending money.

1 Energy

a) Does a lamp use the electrons in the current or the energy they carry to produce light? ______

b) Where does the energy in a circuit come from? ______

2 Complete this table about voltage.

another word for voltage		instrument that measures it	
symbol		joule	
unit			

3 Voltmeters

a) Why are voltmeters connected in parallel and not in series in a circuit?

______

b) Why do you think voltmeters have a high resistance?

______

c) What would happen if they did not have a high resistance?

______

4 Explain what is wrong with these circuit diagrams.

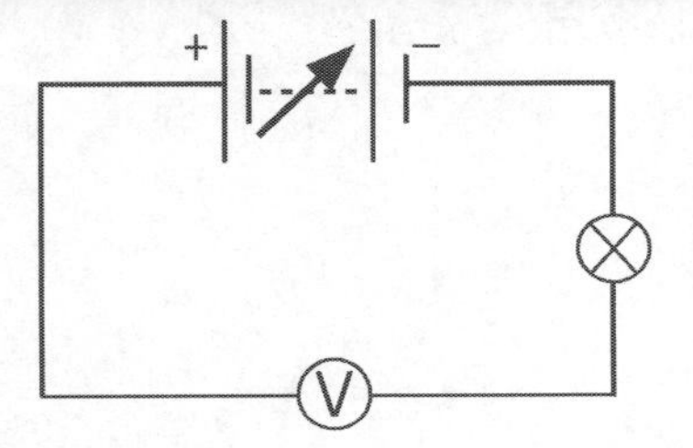

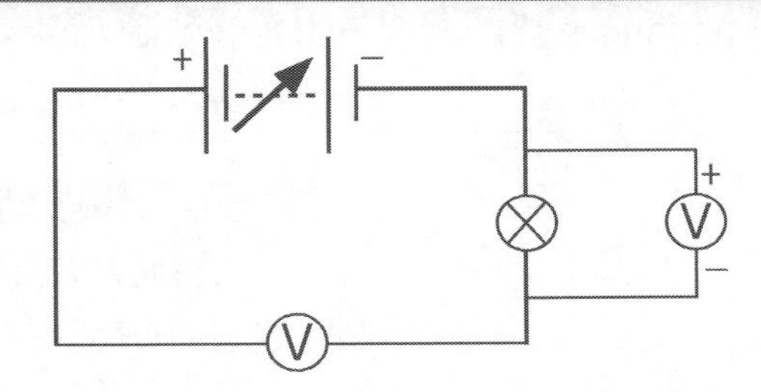

______ ______

5 A student sets up two circuits. Circuit A has two identical lamps in series connected to a 10 V power supply and Circuit B has two lamps identical to those in Circuit A connected in parallel and connected to a 10 V power supply.

a) In which circuit will the lamps be brightest? Explain your answer.

______

______

b) Will adding a third lamp in series in Circuit A make a difference to the brightness of the first two lamps? Explain.

______

______

6 Write down three things you have learned from this unit.

a) ______

b) ______

c) ______

7 Write down anything you need to ask your teacher to explain.

______

UNIT 25 Electric Power

Have you ever wondered why if you replace a 60-watt light bulb with a 100-watt light bulb the light produced is much brighter? After all they are in the same light socket and connected to the same power supply. The answer has to do with how quickly each light bulb transforms electrical energy into light energy. The rate (how quickly) electrical energy is transformed into other forms of energy is called **power**. It is measured in watts. Large amounts of power are measured in kilowatts (kW). There are 1000 watts in a kilowatt.

A 100-watt light bulb transforms 100 joules of energy every second. A 60-watt light bulb transforms only 60 joules of energy every second.

For the same reason, a 1000-watt microwave cooks food faster than a 600-watt microwave.

Labels on appliances show the wattage the appliance operates at as well as the voltage they need to work.

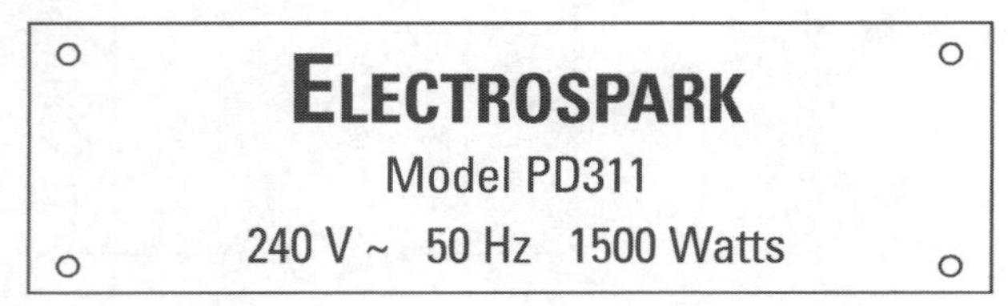

Households in New Zealand usually pay monthly for the electricity they use. On the outside wall of your house you will find a meter box. It contains meters that measure the electricity usage of your home. There are usually two meters: a '24-hour' meter and a 'controlled' meter. The 24-hour meter measures electricity supplied and used 24 hours a day. This is usually the more expensive charge per unit on your power bill. The controlled meter measures the electricity supplied for only part of a 24-hour period (usually at night) and is mainly used for water heating.

The power company can turn this supply off at peak periods to even out the load on the national grid. This is the cheaper charge on your power bill. Households pay for the number of units of electricity they use. One unit is one kilowatt-hour, which is 3.6 million joules of energy.

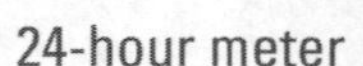
24-hour meter

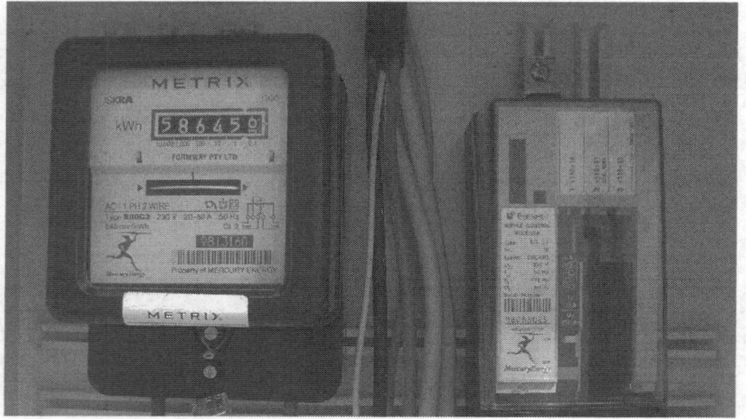

controlled meter

To work out how much energy an appliance uses in one hour, multiply its power (watts) by 3600 (seconds in one hour). For example if you leave a 100-watt light bulb on for one hour, the power it uses is:

100 watts x 3600 seconds = 360,000 joules

The number of units it uses in one hour is:

0.1 kW x 1 hour = 0.1 units

If electricity costs 20 cents per unit, then it costs 20 x 0.1 = 2 cents to leave a 100-watt light bulb on for one hour.

4U2DO

1 In your own words explain what electric power means.

2 Wattage

a) When you are shopping for a microwave, what does the wattage on the label tell you about the microwave?

b) What is the advantage of buying a microwave with a low wattage (think about the power bill)?

3 TV usage

a) How much energy does a 200-watt television use in one hour?

b) How much would the television cost to run if it was left on for one hour?

4 Here is a power bill. Circle how many units this family used in the 36-day period.

Meter reading(s) - For the period

Price plan	This reading	Last reading	Units used
Standard - All Inclusive	57718 (actual)	56725 (actual)	993 kWh

Current account details - For the period

Charge type	Units		Mercury Energy			VECTOR Limited	
Variable usage charge							
Standard - All Inclusive	993 kWh	@	8.83 cents/kWh	$87.68	@	5.38 cents/kWh	$53.42
Daily fixed charge	36 days	@	30.72 cents/day	$11.06	@	11.43 cents/day	$4.11
Metering	36 days	@	16.12 cents/day	$5.80			
EasyPay discount				$3.08cr			
Electricity Commission levy	993 kWh	@	0.17 cents/kWh	$1.69			
Subtotals				$103.15			$57.53
GST at 12.50%				$12.89			$7.19
Totals				$116.04			$64.72
Discount for prompt payment *				$11.61cr			$6.47cr

Total current charges (Mercury Energy plus VECTOR Limited) $180.76

Usage information

Cost per day for the period(s) shown above $5.02/day

5 Write down three things you have learned from this unit.

a)

b)

c)

6 Write down anything you need to ask your teacher to explain.

UNIT 26 Electricity and the Human Body

You get an electric shock when electricity flows through your body. This happens when you become part of an electric circuit, such as when you touch a live wire. You have provided a pathway between the high-voltage live wire and the electrically neutral earth.

The human body's nervous system works using very small electric signals produced by chemical ions. The nervous system controls our breathing, the beating of our heart, the movement of our muscles and many other processes.

An electric current can harm a person because it over-stimulates the nerves that make muscles contract. This can have a number of effects. For example, an electric current can cause the biceps muscle, which makes your arm bend, to contract at the same time as the triceps muscle, which makes your arm straighten. Both muscles pulling on the upper arm bone at the same time but in opposite directions can cause bones to break or become dislocated.

An electric shock can throw the person away from the source of electricity. Contact with the ground can injure the person. Or the person may not be able to let go of the source of the current, so the current passing through the person burns him or her. When an electric current travels across the chest, for example when it goes from one hand to the opposite foot, it can disrupt the signals to the heart causing the heart to fibrillate or shiver wildly. While the heart is doing this, it cannot pump blood. This can lead to a heart attack and death. An electric current can also burn skin and tissue.

Defibrillator

Hospitals and ambulances use a machine called a defibrillator when dealing with a person whose heart has stopped. A defibrillator supplies a brief electric shock to the heart so it returns to a normal rhythm of beating. Defibrillators are made up of two metal paddles or adhesive pads connected by wires to a generator. The paddles deliver the shock through the chest.

Some people who have damaged heart muscle rely on a device called a pacemaker that is surgically placed into the chest near the heart to send an electric impulse to the heart to keep it beating with a normal rhythm.

People do not always die when they get an electric shock. The severity of an electric shock depends on the size of the current (flow of electrons), the voltage (electrical energy) and the resistance of the pathway it follows through the body. Skin has a lower resistance when it is wet than when it is dry. The minimum value for the 'let-go current' is around 10 mA (mA = milliamps = one thousandth of an amp).

Current	Effect on body in dry conditions (one-second contact)
1 mA	Feel a tingling sensation
5 mA	Maximum harmless current
10 mA	Muscles contract – 'can't let go'
100 mA	Heart fibrillates but breathing continues
6 A	Heart fibrillates, breathing stops, burns to body

Voltage	Effect on body in dry conditions
10–12 V	Tingling feeling
15 V	Feel pain
20 V	Severe pain
20–25 V	Can't let go
40–50 V	Death

The voltage of mains electricity in your home wall sockets is 240 V with a current of 10 amps, so a child poking a fork into a socket will receive a severe electric shock.

Here is what you should do if someone suffers an electric shock.

Call out for help – someone else can call 111 while you help the shock victim.
Turn off the source of current if possible or push the person away from the source of current using an insulator (for example wooden stick or pole).
A, B, C: check airway (A) to see if it is clear, check to see if the person is breathing (B), check to see if the heart is beating – circulation (C).
Start mouth-to-mouth resuscitation if the person is not breathing and chest compressions if the heart has stopped beating. This is called CPR – cardio-pulmonary resuscitation. Continue until the person starts breathing, the heart starts beating or professional help arrives.

CPR

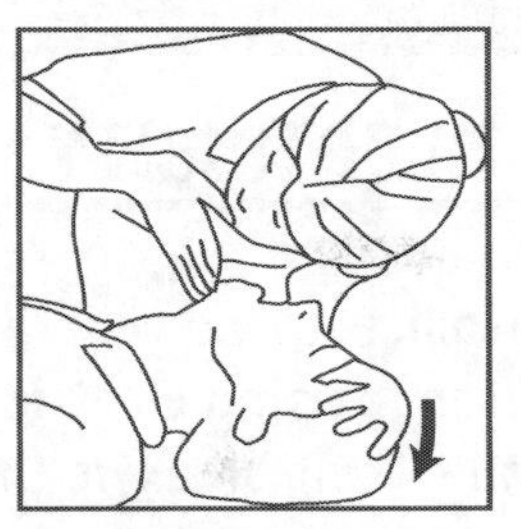

A Check the airway

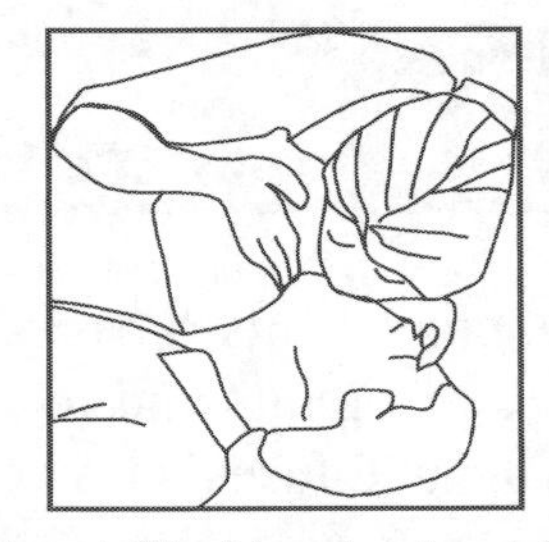

B Tilt head, lift chin, give 2 quick breaths at start, then 15 per minute

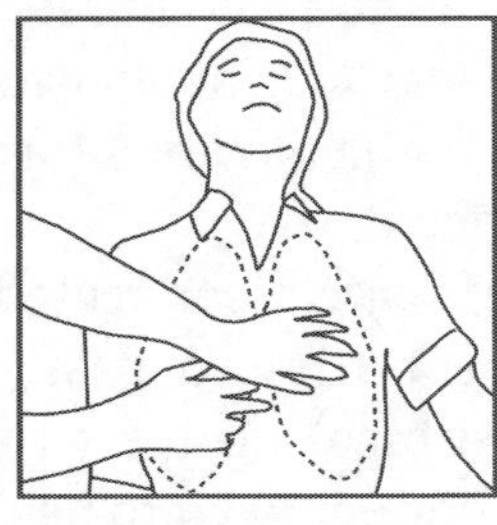

C Position hands in the centre of the chest

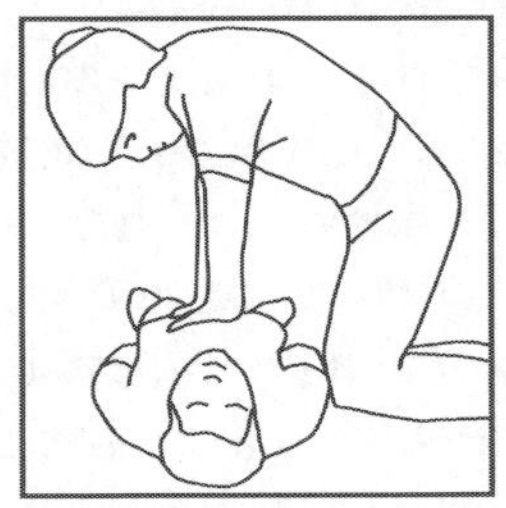

C Push down chest about 5 cm each compression. Try to give 60 compressions per minute

4U2DO

1 What do the human body and electricity have in common?

____

2 Electric shock

a) What are two ways in which an electric shock can harm a person?

____

b) Give one example of how an electric shock can help someone.

____

3 Compare the effects of touching the mains supply with dry hands and with wet hands.

Dry: ________________________________

Wet: ________________________________

4 Shock

a) Why should you call for help when you are dealing with a person who has been shocked?

____

b) What do steps A, B, C stand for?

____

5 Write down three things you have learned from this unit.

a) ________________________________

b) ________________________________

c) ________________________________

6 Write down anything you need to ask your teacher to explain.

____

UNIT 27 SAFETY WITH ELECTRICITY

Fuse wire

A fuse

Circuit breaker

Household circuits have safeguards in place to prevent fires and damage to circuits and to stop people from getting electric shocks. These include fuses (in older homes) and circuit breakers (in newer homes).

A fuse is a thin wire that is a weak link in the circuit. It is designed to melt if too large a current tries to pass through it. This breaks the circuit and prevents a fire starting, appliances being damaged and people getting shocks. The problem with a fuse is that once it 'blows' (melts), you have to replace it.

A circuit breaker is made up of a switch connected to an electromagnet. When too large a current tries to pass through the circuit breaker, the switch is opened breaking the circuit. The advantage of a circuit breaker is that it does not need to be replaced after it has 'tripped out'. Pushing in a button closes the switch and completes the circuit, which resets the circuit breaker.

A house has several circuits connected in parallel. Each circuit is protected by its own fuse or circuit breaker, which you can see in the 'fuse box'. A 5-amp fuse is usually used to protect the lighting circuit in a house, while a 10- or 15-amp fuse protects the wall sockets.

Appliances such as the fridge, microwave and television that use mains electricity have a cord or electric flex and a three-pin plug that you plug into the wall socket. The cord and the plug usually contain three wires.

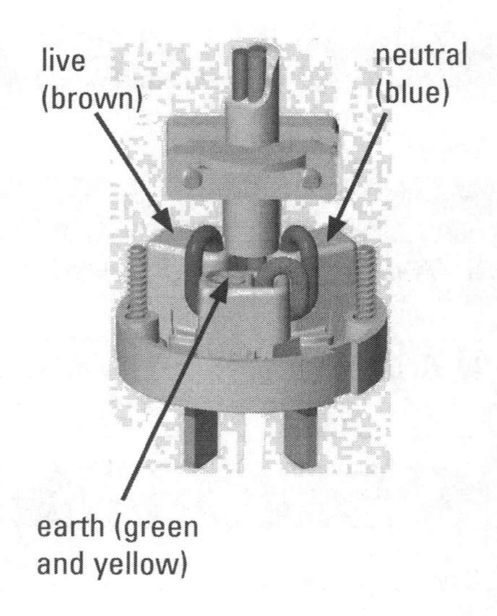

Name of wire	Colour of wire	What the wire does
phase or live	brown	Brings the current into the appliance.
neutral	blue	Completes the circuit and takes the current away.
earth	green and yellow	Safety wire: carries current to earth if a fault occurs.

Where there are only two pins on a plug, the earth wire is missing. This is usually because the plug is double insulated.

The plastic or rubber material around the wires and the cable are insulators and protect people from getting an electric shock. Do not use appliances with bare wires or frayed covering.

When working with electrical equipment outside, having a residual current device reduces the risk of an electric shock.

An RCD compares the current in the live wire with that in the neutral wire. Any difference may indicate a current 'leakage', for example when a person has touched a live component of an

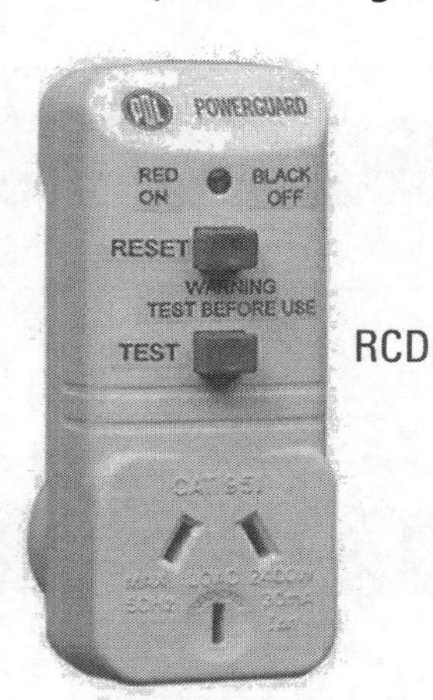

RCD

Never use electrical appliances near water.
Never handle plugs with wet hands.
Never unplug an appliance while it is still on.
Never fly a kite or put up a yacht mast near overhead lines.
Never poke a metal object into a wall socket or an appliance while it is on.

appliance and is getting an electric shock, or a fault in the circuit is letting current take a different pathway. The RCD shown in the diagram 'trips out' or breaks the circuit when a difference of only 30 mA is detected. This breaking of the circuit takes only 30 milliseconds.

An isolating transformer is another piece of equipment used to provide protection when using electric equipment outside or near water. The mains supply comes into the transformer via a connection to the mains supply and the equipment being used is plugged in the other side of the transformer. Inside the transformer, the mains current and the current going to the equipment are separated by a magnetic field. On the side going to the equipment there is no 'earth' connection. You cannot get a shock if there is no pathway for the current to complete the circuit (which it does when it goes to earth) or there is no difference in voltage between the two contacts. This is the reason birds can sit on power lines without getting a shock. Both its feet are at the same voltage and the bird is not 'grounded'.

4U2DO

1 How do fuses and circuit breakers protect you from getting an electric shock in your home?

2 Why do you think the lighting circuit in a house has a smaller fuse in it than the circuit supplying the wall sockets?

3 Wires

a) Name the three wires found in electric cords and plugs.

b) Which wire is the most dangerous and why?

c) What is the purpose of the earth wire?

4 RCD or isolating transformer

a) When should you use an RCD or an isolating transformer?

b) Why should you use them?

5 What are five ways you can prevent getting an electric shock? Try to think of some that are not in this book.

6 Write down three things you have learned from this unit.

a)

b)

c)

7 Write down anything you need to ask your teacher to explain.

UNIT 28 Earth's Structure and Continental Drift

The Earth is a sphere but not a perfectly round one. The circumference around the equator is slightly larger (40,075 km) than the circumference through the poles (40,008 km).

Approximately 71 per cent (or 362,000,000 km^2) of the Earth's surface is covered in water and 29 per cent (or 148,000,000 km^2) is covered by land. Because of the large amount of water on the surface of the Earth, it appears as a 'blue ball' from space.

The Earth is made up of several layers or spheres that are different in structure and composition. Scientists get their knowledge of the inside of the Earth from studying seismic waves, produced by earthquakes, as the waves travel through Earth.

The lithosphere This is the thin, rocky crust and the upper layer of the mantle. The continental crust has dry land on it. On average it is 30 km thick. The oceanic crust forms the sea floors. On average it is only 6 km thick.

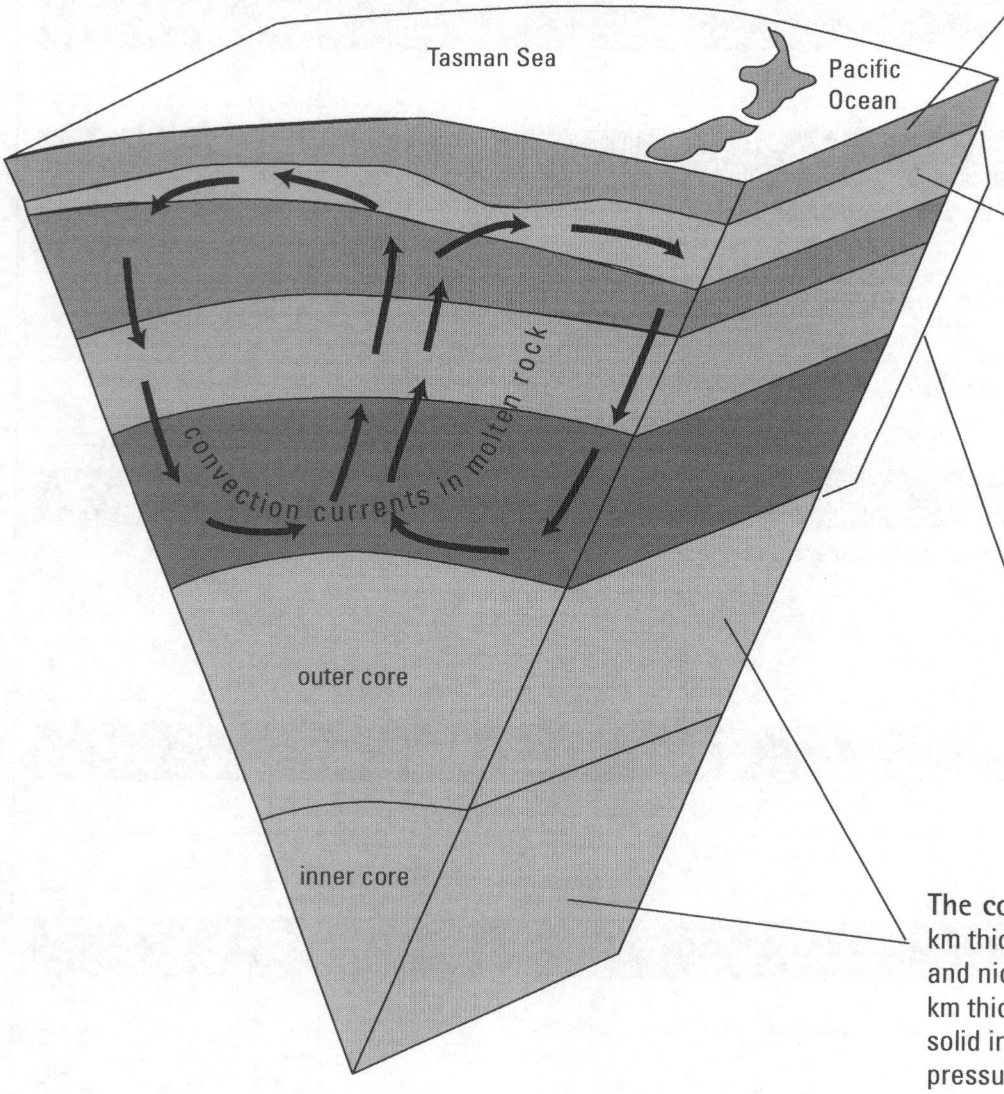

Tasman Sea

Pacific Ocean

The asthenosphere This is the part of the mantle greater than 70 km below the Earth's surface. The temperature is over 1300°C and increases the deeper it goes. These high temperatures make the rock molten. It moves like thick toffee.

convection currents in molten rock

The mantle is a very thick (2900 km) layer between the crust and the core. The upper layer directly under the crust is solid rock with a temperature of about 1000°C.

outer core

inner core

The core The outer core is about 2500 km thick. It is made up of molten iron and nickel. The inner core is about 2600 km thick. It is thought to be made up of solid iron and nickel owing to the huge pressures exerted on it by the mass of the other layers. The temperature at the centre of the Earth may be as much as 5000°C.

Looking at a world map you will notice how the shape of the east coast of South America might fit, like the piece of a jigsaw, next to the west coast of Africa. Scientists made this same observation in the early 1800s. They suggested that when the earth was first formed there was only one landmass, which cracked and broke apart because 'the globe was out of balance'. In the early 1900s a geologist pointed out that fossil evidence showed that in the past South America, central and southern Africa, Madagascar, India and Australia had all shared very similar plant and animal life. It was suggested that they had once been joined together as one continent called Gondwanaland. It was suggested that a second landmass, called Laurasia, had existed in the northern hemisphere.

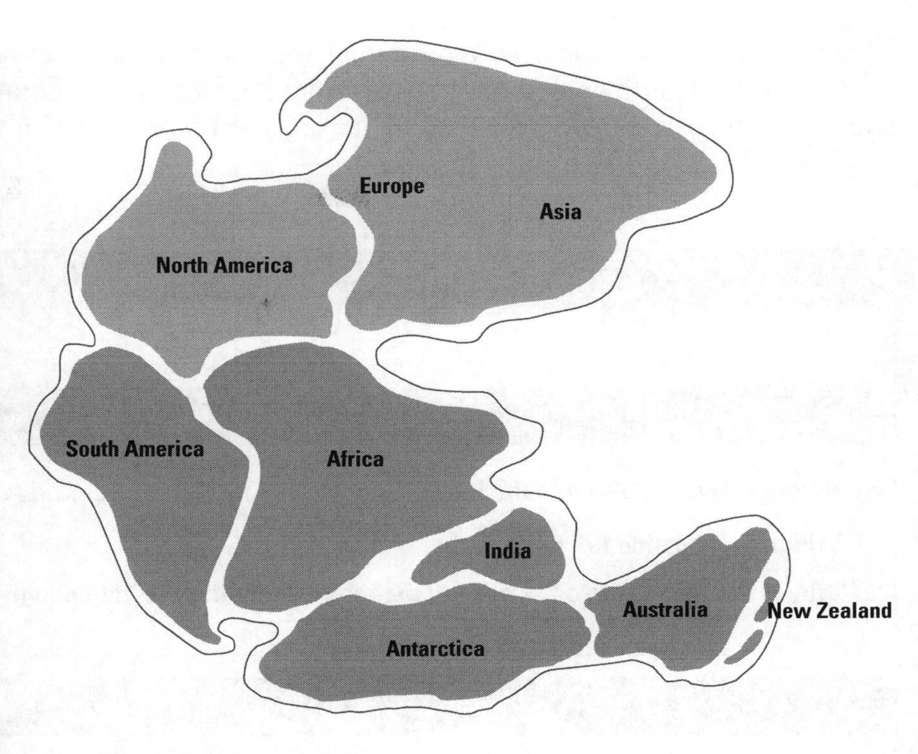

A German meteorologist called Alfred Wegener put forward the theory of continental drift in 1912. This theory suggested that over 200 million years ago all the continents were joined together as one giant continent called Pangaea (meaning 'all the earth'), surrounded by a giant ocean called Panthalassa.

About 200 million years ago, Pangaea broke into two supercontinents called Laurasia and Gondwanaland. These then broke up into smaller continents. They have slowly drifted to their current positions.

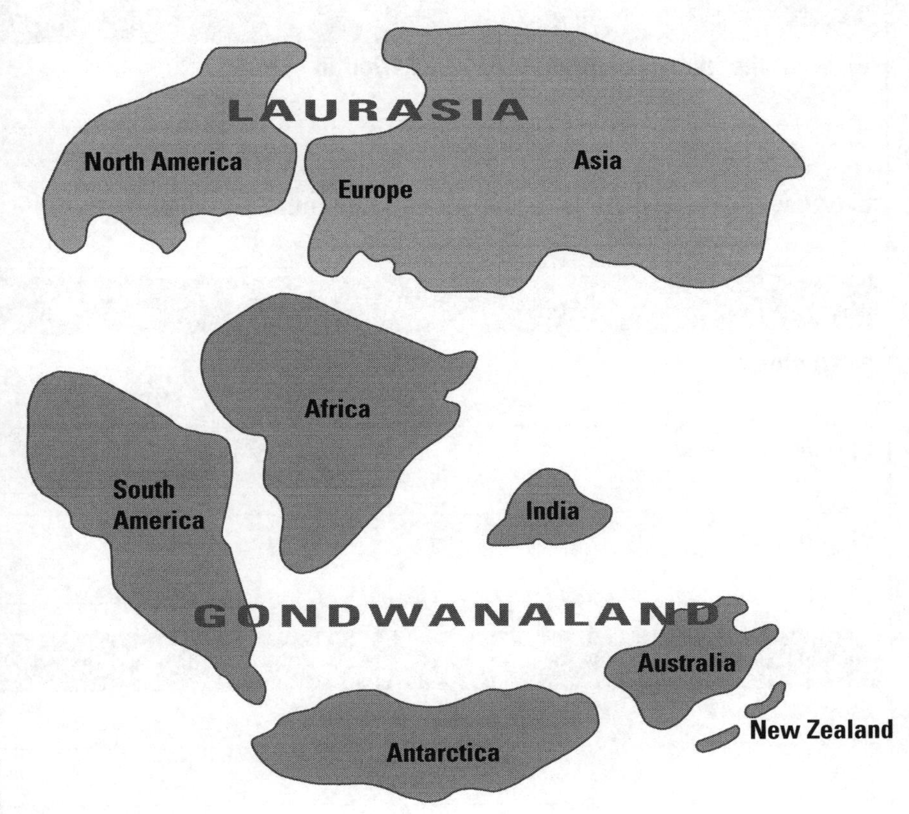

Wegener thought there could be three reasons why the continents moved. These were:

1 The Earth's rotation or spin produced a centrifugal force that pushed the continents away from the poles and towards the equator.
2 The gravitational pull of the moon and the sun affected the continents and moved them in a westerly direction.
3 Gravity pulled some continents to new positions.

Even Wegener himself came to believe that these were not the reasons continents moved. He thought that it might have something to do with convection currents within hot rock inside the earth. In the mid-1900s this idea was developed and became the theory of plate tectonics (see Unit 29).

Scientists are always looking for different ways to apply theories. For example, recent studies of Earth's oldest rocks suggest that the break-up of continents did not start with Pangaea. So now scientists are looking even further back in time.

4 U 2 D O

1 Name two ways that scientists have found out about the inside of the Earth.

2 Think about the structure of the Earth.

a) Where is the Earth's crust thickest?

b) Which layer inside the Earth is the thickest?

c) Why is the inner core solid even though the temperature is hot enough to melt metal?

3 Scientists believe that millions of years ago all the continents were joined together as one giant continent.

a) Name that one giant continent.

b) Name the giant landmass found in the northern hemisphere 150 million years ago.

c) Name the giant landmass found in the southern hemisphere 150 million years ago.

4 In your own words explain

a) what the 'theory of continental drift' tried to explain

b) what evidence there is to show that continental drift is occurring.

5 What do these people study? (Use a dictionary if you are unsure.)

a) geologist

b) meteorologist

c) biologist

6 Write down three things you have learned from this unit.

a)

b)

c)

7 Write down anything you need to ask your teacher to explain.

UNIT 29 PLATE TECTONICS

The theory of plate tectonics developed from the theory of continental drift. It was put forward not only to explain the movement of the continents over time but also to explain the presence of ridges along the ocean floor and why volcanic and earthquake activity occurred mainly in certain areas around the world. This theory is based on the idea that the Earth's continents are embedded in pieces of crust called **tectonic plates**.

There are seven continents: Asia, Africa, North America, South America, Australia, Europe and Antarctica.

Earth's Main Plates

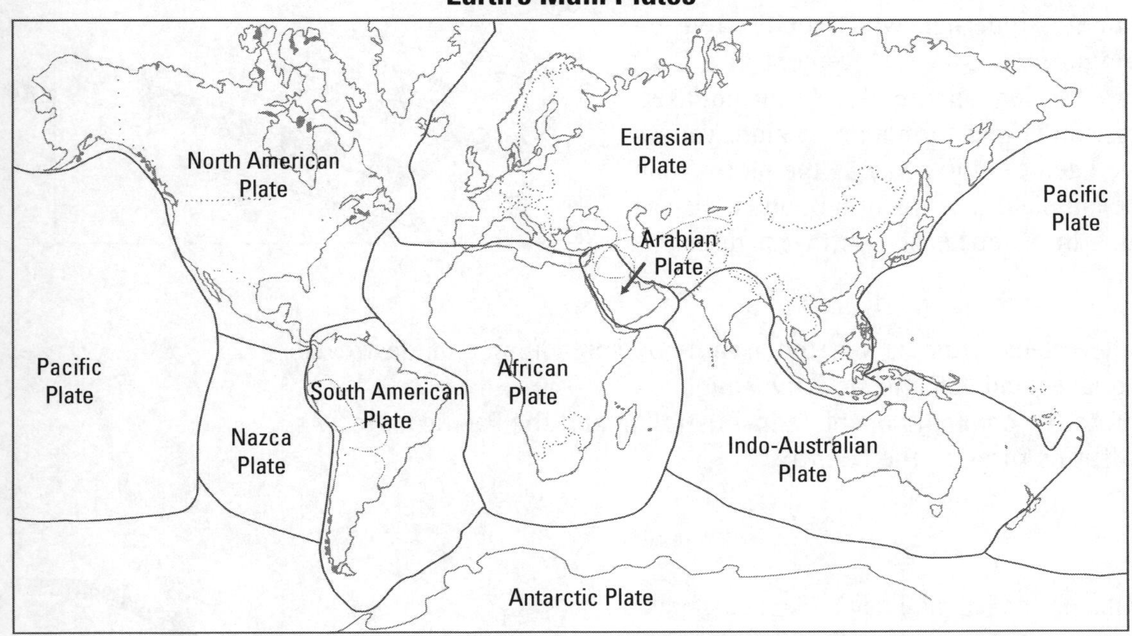

The lithosphere is not one complete layer but is made up of many pieces called tectonic plates that fit together like pieces of a jigsaw (ten large plates and several smaller ones). Most plates consist of continents and sea floor. The Pacific Plate, however, is entirely sea floor. These tectonic plates float on the asthenosphere (molten mantle). They move at speeds of up to 10 cm per year due to convection currents that occur in the molten rock beneath them.

Where one plate joins a neighbouring plate is called a plate boundary. Three things can happen at plate boundaries:

1 At a **divergent** plate boundary the plates are moving apart or 'rifting'. As the plates pull apart, faults develop. The crust between the faults sinks into the asthenoshere where it melts. Hot molten rock or magma then pushes up between the plates. It flows as lava onto the crust surface. There it cools and hardens, forming new crust. This happens most often on the ocean floor. Huge underwater mountain ranges called ocean ridges are formed there. This results in sea floor spreading. As a result the Atlantic Ocean is continually widening. When rifting occurs on land, it creates gaps that water can flow into to form a river, lake or sea.

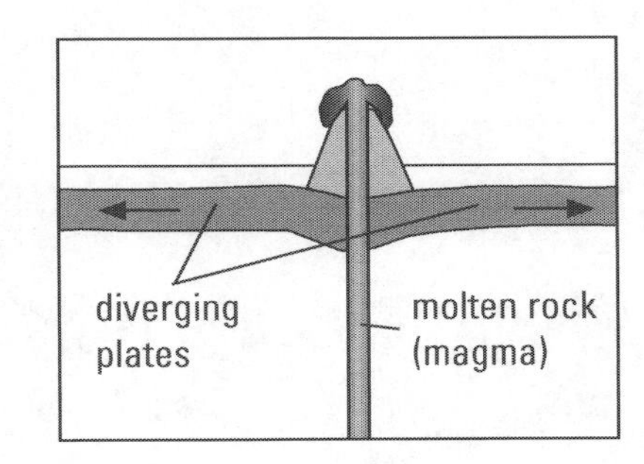

2 At a **convergent** plate boundary plates crash or crunch together. They do this very, very slowly. One plate usually moves down under the other plate. This is called **subduction**. The continental and sea floor crust of the plate moving down is destroyed, as it becomes molten rock in the asthenosphere. When subduction occurs on the ocean floor, it creates huge trenches. These are the deepest parts of the ocean; most are 3–4 km deep. The deepest trench is the Mariana Trench in the Pacific Ocean off the coast of

Japan (11 km deep). Subduction at boundaries around the Pacific Plate is the reason why the Pacific Ocean is shrinking.

Sometimes when boundaries crash together, the front ends of the plates bend and fold together and mountains are formed. For example the Himalayan Mountains were formed when the Indo-Australian Plate, carrying India, pushed beneath the Eurasian Plate, and they continue to grow in height as the plates push against each other. Another example is where the oceanic Nazca Plate crashed into the continent of South America. This formed the Andes Mountains and the deep trench off the coast in the Pacific Ocean.

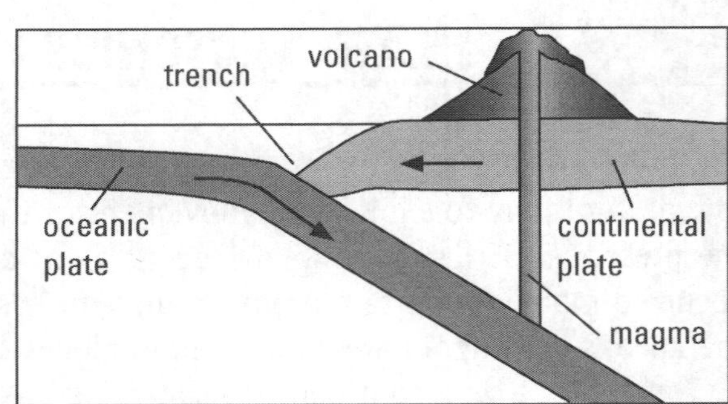

3 At **transform** boundaries, plates slide past one another. The plates are pushed tightly together, which builds tension at the boundary forming fault zones. Movement at a fault releases the built-up tension and produces earthquakes. Transform boundaries sometimes form long, straight valleys where the rocks have been ground away as the plates slide past each other. An example of a transform boundary is the Alpine Fault that runs in a northeast–southwest direction through New Zealand.

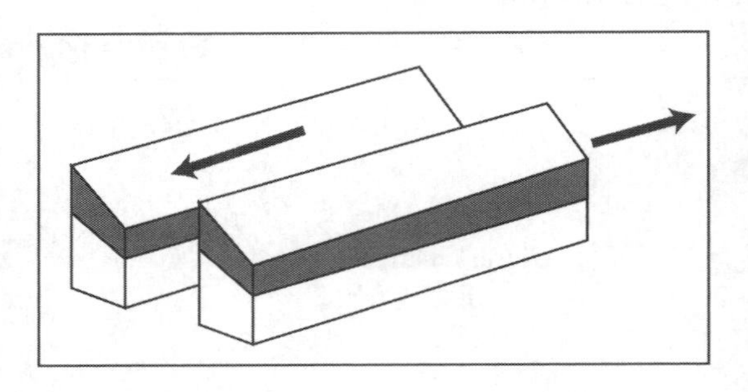

Where tectonic plates meet, crust is created (sea floor spreading) and destroyed (subduction) and earthquakes and volcanic activity is high.

New Zealand sits across the boundary of the Indo-Australian and the Pacific plates. As a result, the Alpine Fault runs through the country.

Tonga-Kermadec Trench

Pacific Plate

Hikurangi Trough

Indo-Australian Plate

Puysegur Trench

1 To the northeast, the Pacific Plate subducts under the Indo-Australian Plate forming the Tonga-Kermadec Trench. The Pacific Plate is also moving in a southwesterly direction.

2 To the southwest, the Indo-Australian Plate subducts under the Pacific Plate forming the Puysegur Trench. The Indo-Australian Plate is also moving in a northeasterly direction.

3 The Alpine Fault is about 400 km long and runs through the South Island. It seems to connect the subduction zones at either end of New Zealand. Land around the fault is moving at a rate of 35 mm per year. Fossil and rock evidence suggests that over the past 25 million years there has been a land shift along the fault line of 450 km. This movement happens in jerks, felt as earthquakes, rather than a smooth motion. The plates last moved in a major earthquake in 1717.

The Ring of Fire is a belt (40,000 km long) that encircles the Pacific Ocean. Along this belt, where many convergent boundaries occur, are 350 volcanoes (half of the Earth's active volcanoes). It is here that 80 per cent of the Earth's major earthquakes occur.

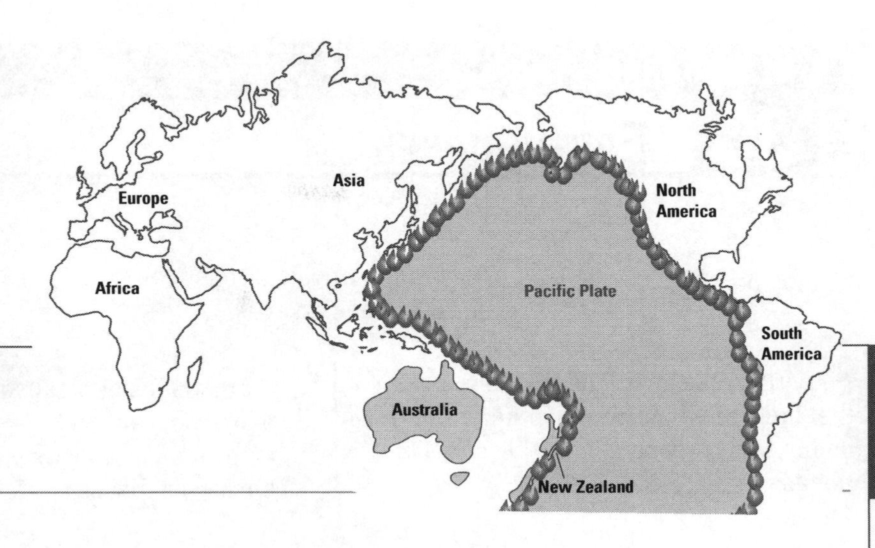

4U2DO

1 What do the following terms mean?

a) tectonic plate

b) convergent

c) divergent

d) subduction

2 Large pieces of the earth's crust are called tectonic plates. Name:

a) the two tectonic plates New Zealand sits across

b) two other major tectonic plates

c) a plate that has no continents on it.

3 Tectonic plates can move apart, come together or slide past each other. Describe, using a named example, what happens when plates:

a) move apart

b) come together

c) slide past each other.

4 Look at the diagram showing the tectonic plates and the one showing the Ring of Fire. What link is there between the two diagrams?

5 Write down three things you have learned from this unit.

a)

b)

c)

6 Write down anything you need to ask your teacher to explain.

UNIT 30 EARTHQUAKES

A vibration travels through the earth as shock waves.

The **focus** or hypocentre is the point inside the earth where earthquakes originate. Most are less than 70 km deep but the deepest that occur at subduction boundaries can be up to 700 km deep.

An earthquake is produced when rocks that have been pushed together and are under great pressure suddenly move to release the stress.

Earthquakes can cause tsunamis, avalanches and landslides as well as property damage and loss of life.

Most often, earthquakes are a result of tectonic plates moving but can be caused by volcanic eruptions, underground explosions and meteor impacts on earth.

Over 300 million earthquakes occur in the world every year; that's one every 11 seconds! New Zealand has about 10,000 per year but only around 200 are big enough to be felt.

The point on the surface of the earth directly above the focus is called the epicentre. The strongest shaking is felt at the **epicentre**.

Ruamoko was the youngest child of Papatuanuku (Earth mother) and Ranginui (sky father). One version of this tradition says Ruamoko was an unborn child. He was pushed into the underworld when the older children forced Papa and Rangi apart. He became the god of earthquakes and volcanoes. Whenever he is restless, the Earth shakes and rumbles.

Cracks and fractures in weak areas of the Earth's crust are called **faults**. Major faults in New Zealand include the Alpine Fault, Wairau Fault, Wellington Fault, Wairarapa Fault and the Edgecumbe Fault.

They are usually below ground but some, such as the Wairarapa Fault in the lower North Island, can be seen on the surface. Faults can be tens or hundreds of kilometres deep and hundreds or thousands of kilometres long. Faults break up the ground into big blocks. Some faults occur at the boundaries of plates but others occur in the middle of plates. The rock on either side of a fault is under great pressure owing to forces in the Earth's crust. Eventually the pressure has to be released. When this happens, the rock on either side moves suddenly and an earthquake occurs.

There are three basic movements that can occur at faults:

1 In a **normal fault** the fracture in the rock slopes downward and the rock on each side moves up and down. These faults occur where the crust is being pulled apart, for example at divergent plate boundaries.

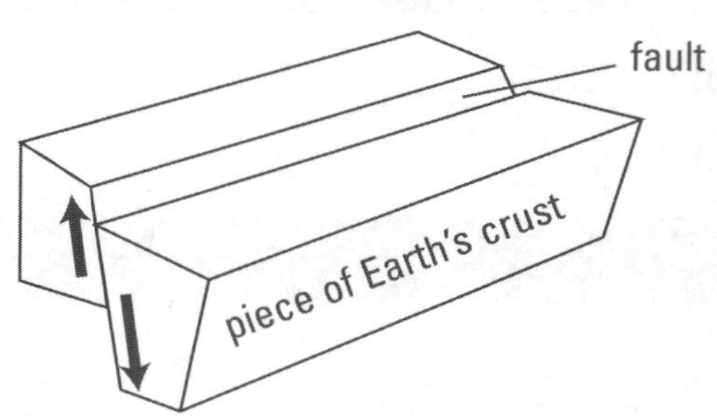

2 A **reverse fault** is similar to a normal fault but it occurs where the rock on each side of the fault is being squeezed together, for example at convergent plate boundaries.

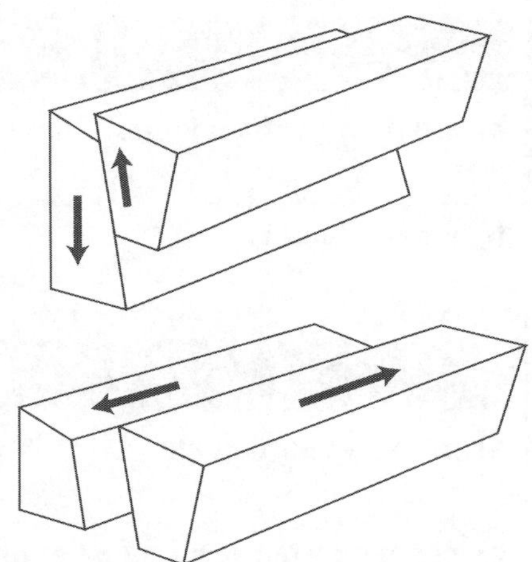

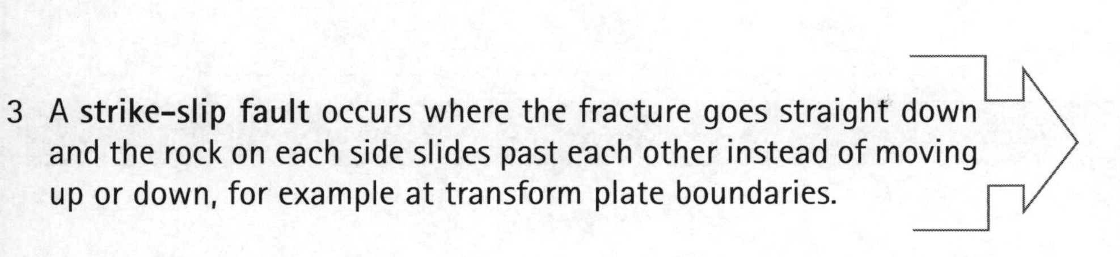

3 A **strike-slip fault** occurs where the fracture goes straight down and the rock on each side slides past each other instead of moving up or down, for example at transform plate boundaries.

The sudden break or shift in rock that occurs at fault lines produces seismic waves that spread out from the earthquake focus in all directions.

Body waves move through the inner part of the Earth. There are two types of body waves:

Primary or **P-waves** are compression (push-pull) waves that move rock back and forth in the direction the wave is travelling. They can travel at speeds of 6 km per second near the surface and 10 km per second near the core.

Secondary or **S-waves** are shear waves. They cause the rock to move side to side at right angles to the direction that the wave is travelling. This rolling motion is what causes damage to buildings. These waves can travel at speeds of 3 km per second near the surface and 8 km per second near the core.

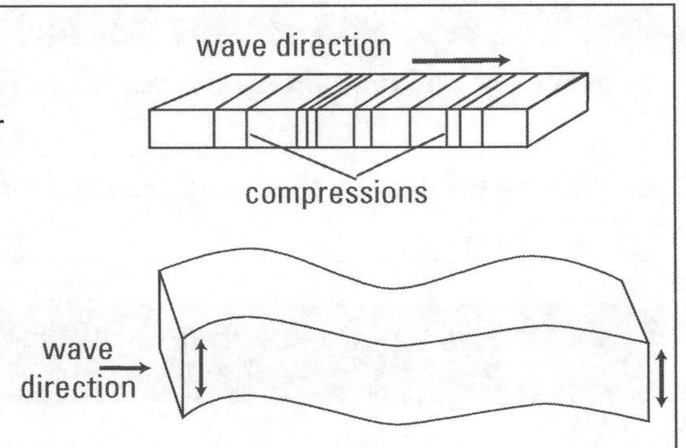

Surface waves travel over the Earth's surface and cause damage to buildings because of the way they make the ground move. There are two types of surface waves:

Rayleigh waves roll like ocean waves at speeds of about 3.5 km per second.

Love waves move the ground from side to side and travel at speeds of about 4.5 km per second.

A seismograph is a device used to measure earthquakes. It records the motion of the ground. Data from seismographs in at least three different locations is used to pinpoint the epicentre of an earthquake. There are three different scales used to measure the magnitude of an earthquake:

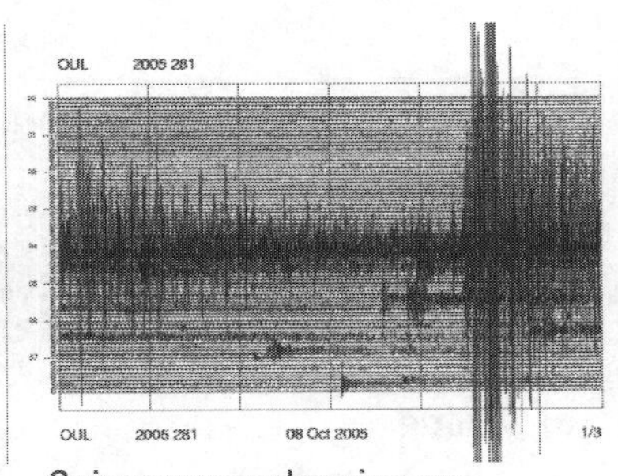

Seismogram showing an earthquake

The **Richter scale** ranges from 0 to 10. It is a measure of the size of the ground motion caused by a quake. An earthquake of magnitude 2 can just be felt while a quake of 6 or higher damages buildings. This scale is logarithmic; each number is ten times larger than the number before it. One of the largest recorded earthquakes was off the coast of Sumatra in 2004. Experts are still discussing its exact measurement – perhaps 8.5–9 on the Richter scale and 9–9.3 on the Moment Magnitude scale. The largest New Zealand earthquake (Wairarapa 1855) measured 8.2 on the Richter scale.

The **Moment Magnitude scale** measures the total energy released by an earthquake and is more accurate than the Richter scale for large earthquakes. An increase of one step on this scale corresponds to a $10^{1.5}$ increase in energy released. An increase of two means 10^3 or 1000 times more energy released. The largest recorded earthquake measured 9.5 on this scale (the Chile earthquake of 1960) measured 9.5 on this scale.

The **Mercalli Intensity scale** ranges from I to XII (it uses Roman numerals). It measures the effect on people, buildings and the environment. On this scale a IV feels like the vibration of a heavy truck driving past. A VII will move furniture and damage buildings not designed to withstand earthquakes. A X causes severe damage to buildings and bridges.

4U2DO

1 In your own words describe:

a) what an earthquake is.

b) what a fault is.

2 List two things that can

a) cause an earthquake

b) happen as the result of an earthquake

c) happen to the land on either side of a fault when it suddenly moves.

3 Name two

a) New Zealand fault lines near where you live

b) major earthquakes that you have read about or heard about.

4 On the outline of the Earth, draw and label: the focus, the epicentre, the seismic waves.

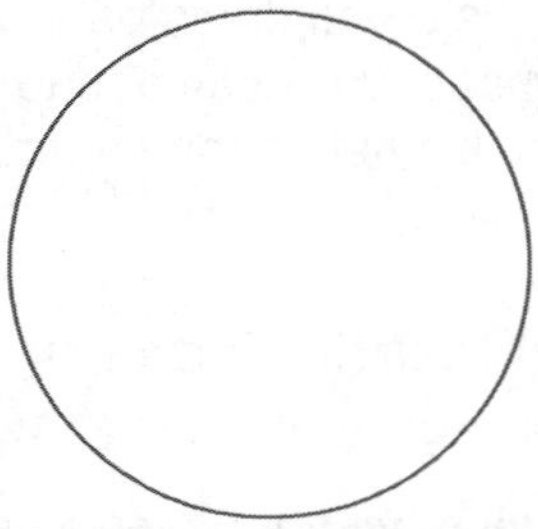

5 Name two differences between

a) P-waves and S-waves

b) P and S-waves and Rayleigh and Love waves.

6 Explain what each of these scales tells you about an earthquake.

a) Richter

b) Mercalli Intensity

c) Moment Magnitude

7 Write down three things you have learned from this unit.

a)

b)

c)

8 Write down anything you need to ask your teacher to explain.

UNIT 31 VOLCANOES

The word volcano comes from the word 'Vulcan', the name of the Roman God of Fire.

A volcano is an opening in the Earth's crust. This can be like a mountain or simply a crack in the ocean floor.

Steam, rocks, ash and lava can be forced out through these openings, sometimes in very violent eruptions.

Most volcanoes occur along the boundaries of tectonic plates although there are some 'hotspots' that occur in the middle of plates.

A volcano is made from magma (molten rock, volcanic gases and solid particles that are found inside the earth).

Volcanoes are different shapes and sizes owing to the magma they are made of.

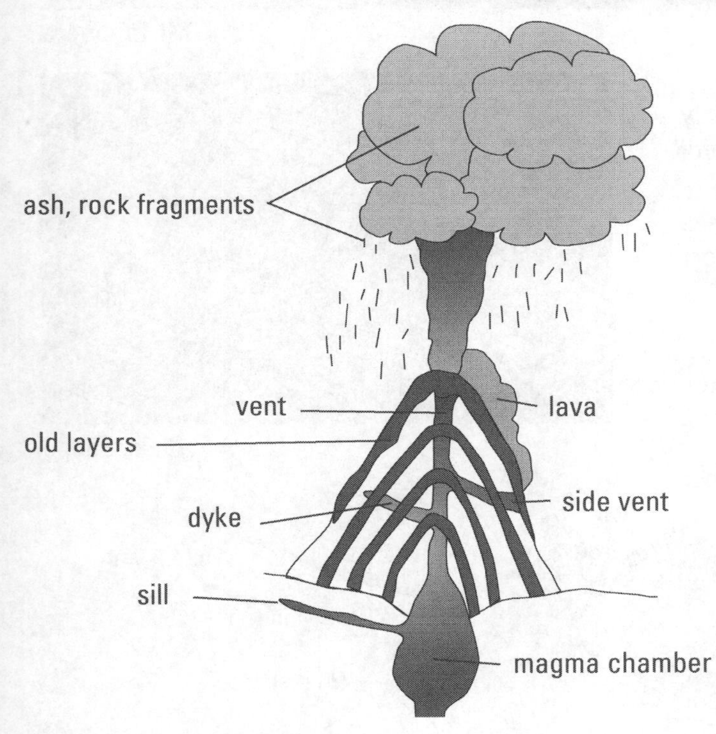

Maori tradition says that the tohunga Ngatoro-i-rangi was caught in a blizzard on Tongariro. With him was Auruhoe. One version says she was a slave, another says she was his wife. Ngatoro-i-rangi called to his sisters in Hawaiki to send him fire. The sacred fire they sent burst out at White Island, Rotorua and Taupo and then made the cold mountains sizzle. One version says Auruhoe died in the blizzard, another says Ngatoro-i-rangi sacrificed her to the gods.

The amount of a mineral called silica (silicon dioxide) in magma determines how easily the magma flows. Magma with lots of silica is sticky like toffee so it flows slowly. Magma with little silica is thin and flows like runny honey. Once magma comes out onto the Earth's surface it is called lava. Lava may be as hot as 1000°C.

Magma also contains dissolved gases such as water vapour, carbon dioxide and sulfuric gases that are under a great deal of pressure, a bit like the gas in a can of soft drink. As the gases rise up towards the Earth's surface, they expand. This makes the magma swell up. Eventually the gas bubbles burst out from inside the Earth in a volcanic eruption. The stickier the magma, the bigger the eruption but the less often they occur.

When cold water and magma meet, the results are even more spectacular as the force resulting from superheated steam blasts away huge amounts of rock.

There are three main types of volcanoes; all are present in New Zealand.

1 Shield volcanoes, for example Rangitoto and the ancient volcanoes in the Christchurch and Dunedin areas. They
- are broad, dome or shield shaped with gently sloping sides
- are made from runny basalt magma where gases escape easily
- have layers of lava flowing slowly and building up the volcano over time.

Rangitoto

2 Cone volcanoes, for example Mt Taranaki, Mt Ruapehu, Mt Ngauruhoe, Mt Tongariro and White Island. They
- are cone shaped with steep sides
- are made of more sticky andesite magma where trapped gases explode breaking the magma into equal amounts of ash and lava
- have layers of lava and ash building up the volcanic cone
- have pumice forming from frothy lava in the vent.

Mt Taranaki

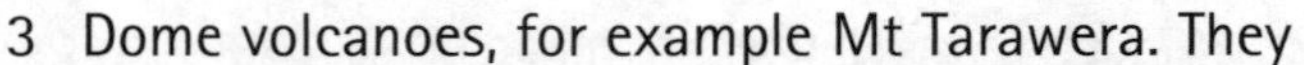

3 Dome volcanoes, for example Mt Tarawera. They
- have lava piling up just above the vent to form domes
- are made of the slowest flowing rhyolite magma that has lost its gas content
- occasionally erupt violently.

Mt Tarawera

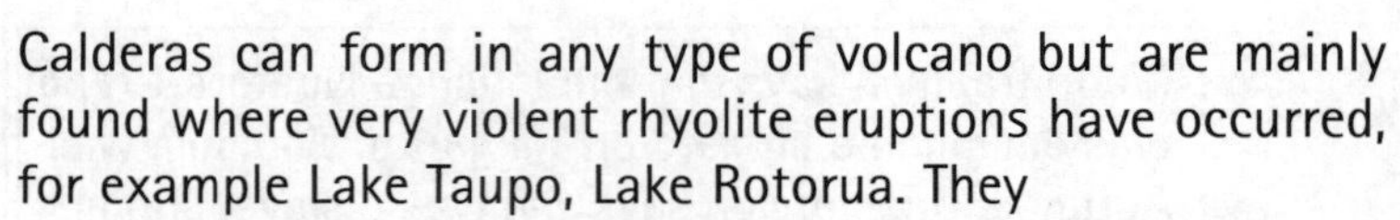

Calderas can form in any type of volcano but are mainly found where very violent rhyolite eruptions have occurred, for example Lake Taupo, Lake Rotorua. They
- are formed when rock that surrounds the vent collapses because the magma chamber is empty so cannot support the weight above
- often fill with water to form lakes.

Lake Taupo

New Zealand has many volcanoes because of its position above the edge of two tectonic plates. Most volcanoes in the South Island are extinct. This means they have not erupted since the beginning of recorded history.

Most of New Zealand's active volcanoes are found in the Taupo Volcanic Zone in the central North Island.

This area is 20–40km wide and 240 km long and extends from Ohakune to White Island in the Bay of Plenty.

The largest eruption in this area was about AD 186. About 100 cubic kilometres of rock was blown into the sky reaching heights of 50 km. The volcanic vent responsible for this eruption lies beneath the northeast corner of Lake Taupo. The lake itself formed in the caldera that resulted from the eruption.

More recently Mt Ruapehu erupted in 1995.

White Island, 50 km off the coast of Whakatane, is the most frequently active volcano in New Zealand. It is the tip of a volcano that is the size of Mt Tongariro. Its many vents release a poisonous, acidic mix of gases and steam.

1 In your own words explain

a) what a volcano is

b) why volcanoes occur mainly where tectonic plates meet

c) what effect the mineral silica has on magma.

2 Without looking back at the text, draw and label the parts of a volcano.

3 For each photo below, say what type of volcano it is and name one example found in New Zealand.

Type:	Type:	Type:
Example:	Example:	Example:

4 Name four volcanoes found in the Taupo volcanic region.

a) ____________ **b)** ____________

c) ____________ **d)** ____________

5 Write down three things you have learned from this unit.

a) ____________

b) ____________

c) ____________

6 Write down anything you need to ask your teacher to explain.

UNIT 32 WEATHERING AND EROSION

The surface of the Earth takes a constant battering from environmental factors such as radiant heat, wind, rain and snow. Plant roots, bacteria and fungi also play a part in attacking the exposed rocky crust. The combination of these actions along with the effects of earthquakes and volcanoes means the face of the Earth is always changing. Sometimes this change happens quickly but often it is very, very slow.

The term **weathering** refers to the breaking down of rocks on the Earth's surface. Rocks can be broken down in three ways:

MECHANICAL BREAKDOWN

1 As a result of heating and cooling, rocks expand and contract. This can create stress cracks in the rock and eventually the rock breaks into smaller pieces.
2 Water in cracks expands when it freezes. This forces the cracks wider and helps break up the rock.
3 Wind slowly wears away at rock surfaces. If the wind carries dust and grit, it acts like a waterblaster hitting the rock surface.

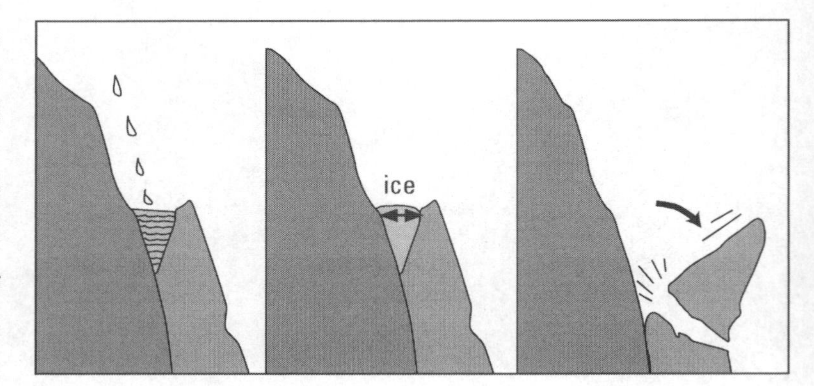

CHEMICAL BREAKDOWN

1 Rainwater can dissolve some minerals in the rock.
2 Carbon dioxide dissolved in rain forms acid rain. It reacts with rocks such as limestone and marble that contain the mineral calcite. This is how limestone caves are formed.
3 Oxygen combines with iron-bearing silicate minerals causing oxidation. This can give rocks a red, orange or brown colour.

BIOLOGICAL BREAKDOWN

1 Plant roots grow into the cracks in rock surfaces. As they grow they widen, splitting the rock into smaller pieces.
2 The actions of lichens, fungi and bacteria wear away rock surfaces.
3 Bird droppings are acidic. This reacts with the calcite in limestone and marble.

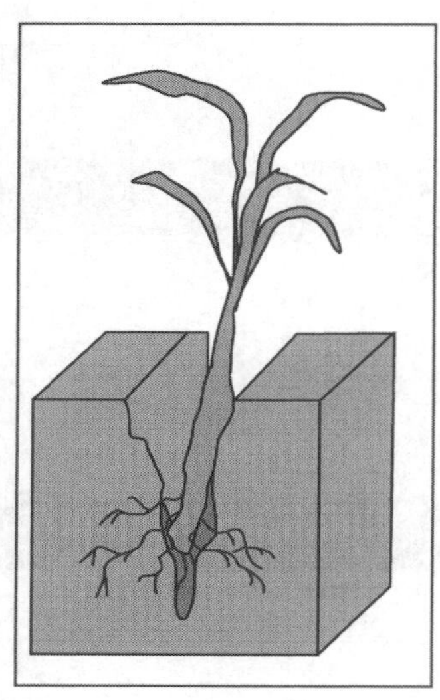

The term **erosion** refers to the transportation of broken-down rock from one place to another. This can be carried out by wind, water and ice.

Wind picks up tiny mineral particles called **loess** (pronounced *lerss* or *lo-iss*) and deposits it on hillsides that face the prevailing wind. Loess forms fertile topsoil. In desert areas wind can completely change the landscape from one day to the next as sand dunes disappear from one area and appear in another area.

Desert landscape

Ice glaciers are frozen rivers of ice. They flow down mountains because of the pull of gravity. They gouge out U-shaped valleys as they go. This is because they rub against the rock of the valley sides and floor, breaking rock up and carrying it away. When the glacier retreats back up the valley, it leaves behind piles of rock called a **moraine**.

A hanging valley at Milford Sound

Flowing water carries dissolved minerals, small particles such as mud and sand and vegetation. It carries them away to be deposited elsewhere. When rivers flow down mountains, they form V-shaped valleys. The current tumbles along rock pieces so their edges become rounded. When the flow of the river decreases, it can leave behind a fan-shaped deposit of gravel, pebbles, sand and silt called an **alluvial fan**, or sand and gravel bars to form a braided river. Flooded rivers leave behind layers of mud once they recede.

Franz Josef Glacier

Waves pounding against cliff faces break up rock into smaller pieces that are carried away by the currents. Waves also drag sand from beaches and deposit it on the sea floor.

The material obtained from weathering and erosion such as sand, mud and pebbles form layers that are eventually compacted and cemented together to form sedimentary rocks. The limestone that forms the Waitomo Caves is an example of sedimentary rock.

Weathering and erosion are responsible for some of New Zealand's tourist attractions such as the Punakaiki Rocks on the west coast of the South Island, the hanging valleys in Milford Sound, and the Waitomo Caves in the central North Island.

Weathering and erosion keep the Southern Alps at an almost constant height despite the fact that the Pacific and Indo-Australian plates are pushing together building them up.

Punakaiki Rocks

4U2DO

1 In your own words explain the meaning of

a) weathering

b) erosion.

2 Below each diagram showing a type of weathering write if it shows chemical, mechanical or biological weathering.

	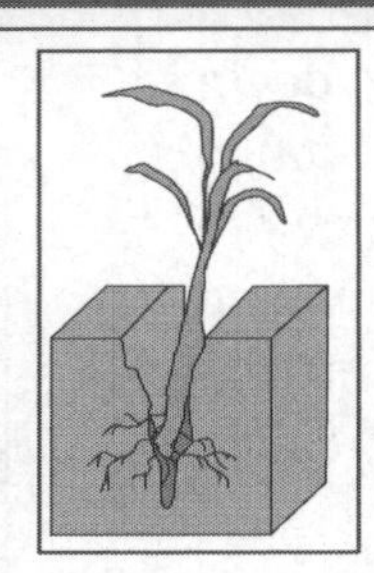	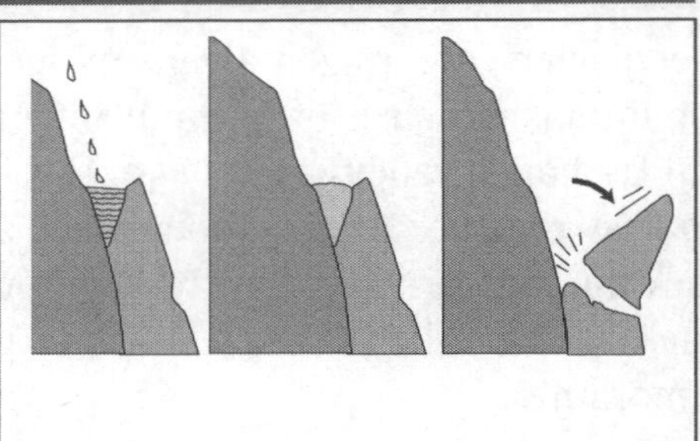

3 In your own words explain these observations.

a) Valleys formed from glaciers are U-shaped.

b) Stones carried by rivers are smooth and rounded.

c) Concrete around large trees is often cracked.

d) Hillsides planted with trees have fewer slips than bare hills.

4 Use an atlas to find where tourist attractions Waitomo Caves, Punikaiki Rocks and Milford Sound are and then mark them on the outline of New Zealand.

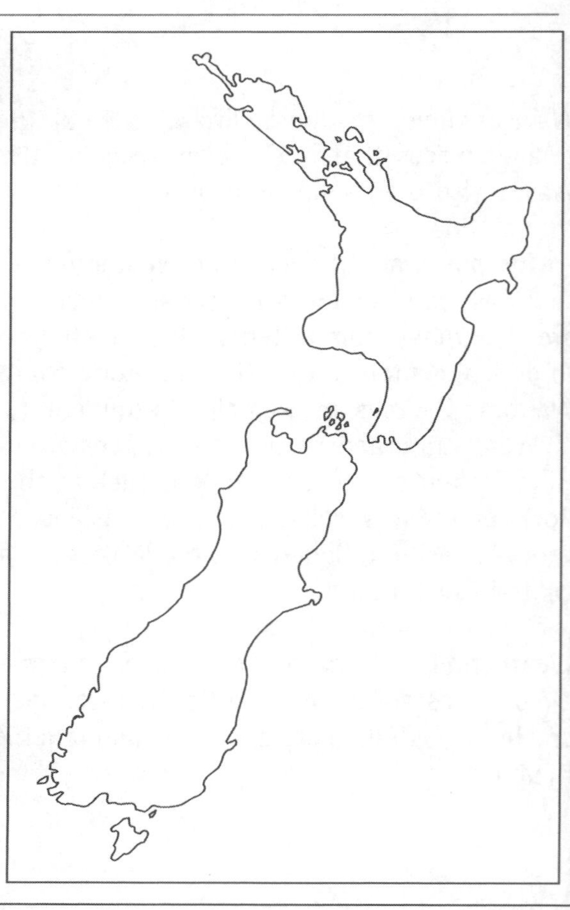

5 Write down three things you have learned from this unit.

a)

b)

c)

6 Write down anything you need to ask your teacher to explain.

UNIT 33 New Zealand's Geological History

The oldest rocks on Earth have been dated (using radioactive decay of uranium and carbon) at just over four billion years old. The oldest rocks in New Zealand, found on the west coast of the South Island, are around 500 million years old, so New Zealand is young in geological terms.

The known history of Earth has been divided up into geological time units. There are four eons, with the first three together lasting four billion years. The last eon, when life became plentiful, is divided into smaller units called eras. Eras are further divided into periods based on fossil evidence of the major plant and animal life of the time found in rocks. It is during some of these latter eras that New Zealand had its beginnings.

The islands that make up New Zealand as we know it today are very different in size and shape to what they were millions of years ago. The New Zealand landmass, along with Australia and Antarctica, began as part of the eastern edge of Gondwanaland. Some of our oldest rocks are found on the West Coast of the South Island.

Much of the rest of time 'New Zealand' was part of the sea floor lying off the eastern margin of Gondwanaland. Amongst this collection of sediments was a series of volcanic islands. Sedimentary material was eroded off Gondwanaland and carried out to sea before being deposited on the sea floor where it built up to form a landmass much larger than New Zealand is today. This land mass was uplifted and folded to form mountains, only to be slowly eroded away again.

About 85 million years ago a spreading margin began to form, owing to divergent tectonic plates moving apart. New Zealand began its separation from Gondwanaland. The Tasman Sea developed as a result of sea floor spreading.

Maori mythology says Maui fished up Te Ika a Maui (the fish of Maui), which became the North Island of New Zealand, from the sea. His brothers gutted it and cut it up before Maui had made an offering to the gods. Offended, the gods made the 'land fish' twist and squirm. These movements formed the mountains and the rugged coastline. (The South Island was Te Waka o Maui, the canoe of Maui.)

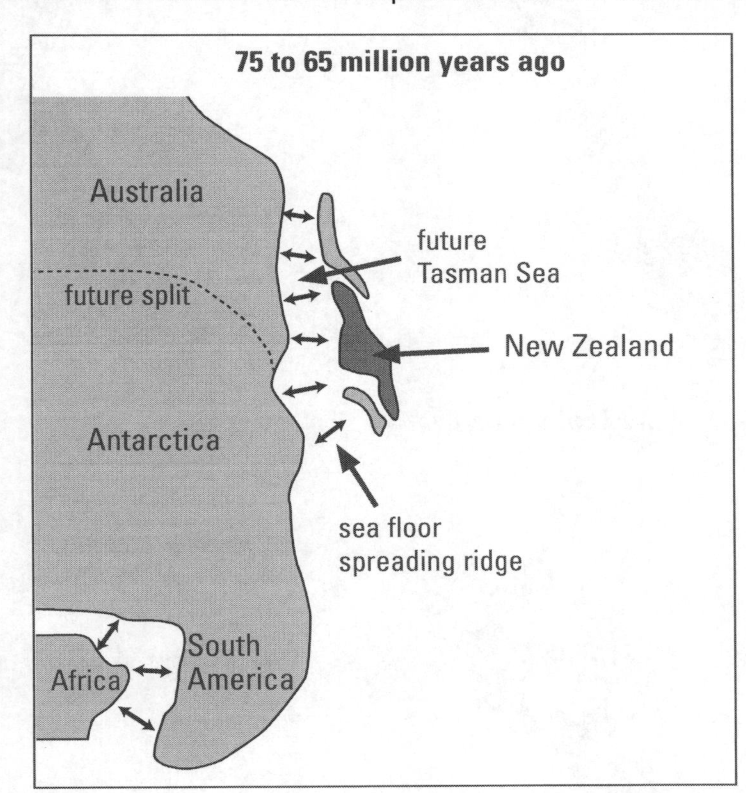

At the same time as the continental crust that carried New Zealand moved away from Gondwanaland, erosion of the landmass continued once again and by 35 million years ago most of New Zealand was under water again.

45 million years ago

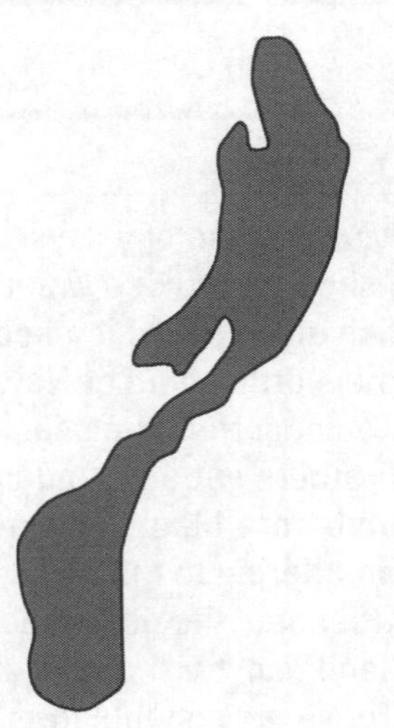

35 million years ago

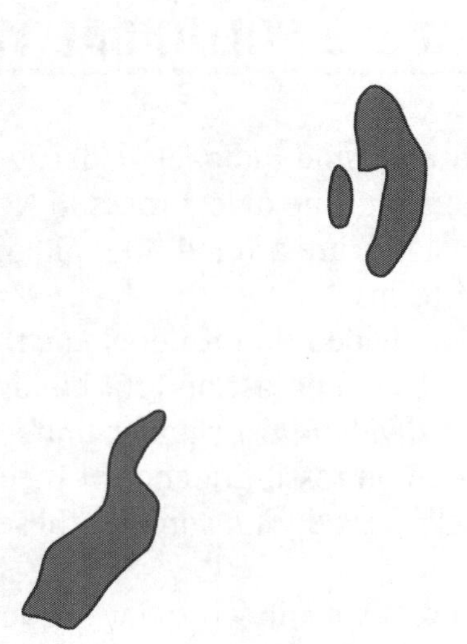

20 million years ago

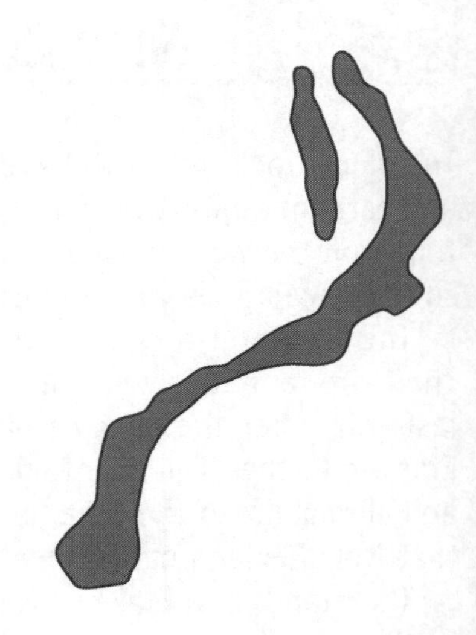

Renewed tectonic activity about 25 million years ago led to a subduction zone forming east of New Zealand as the Pacific Plate collided with the Australian Plate. This process continued through to the present day. During this time the major mountain ranges of both islands were uplifted. Volcanic activity began again and New Zealand began taking its modern shape.

10 million years ago

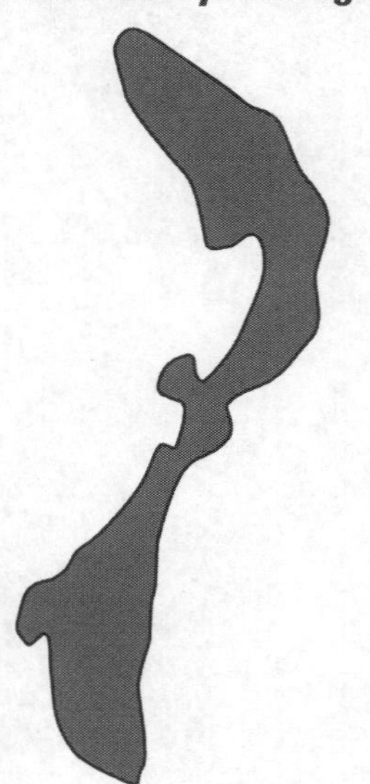

8 million years ago

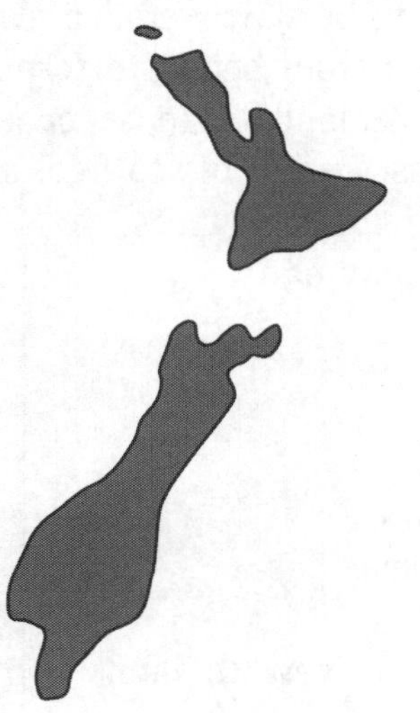

2.4 million to 10,000 years ago

New Zealand today

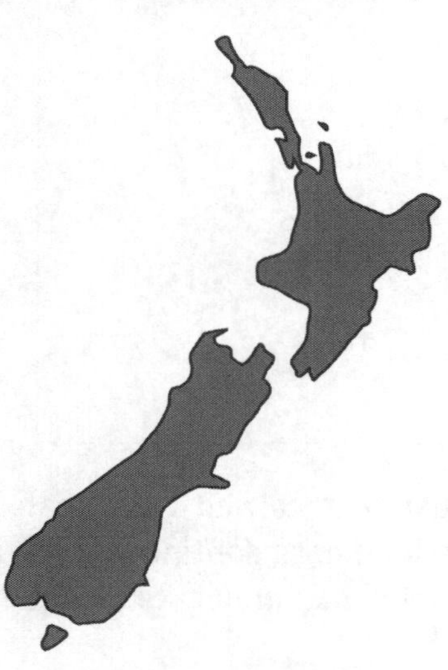

4U2DO

1 Based on rock evidence, at least how old is

a) the Earth?

b) New Zealand?

2 Use the diagrams in this unit to answer the following.

a) How long ago did New Zealand begin to split away from Gondwanaland?

b) When were the North Island and the South Island last joined together?

c) When did New Zealand form an archipelago (a group of many small islands)?

3 In the space below, draw an outline of what New Zealand might look like in another 15 million years.

4 Write down three things you have learned from this unit.

a)

b)

c)

5 Write down anything you need to ask your teacher to explain.

SCIENCE GLOSSARY

A

acid – a compound that contains hydrogen ions and when dissolved in water produces hydrogen ions; has a pH of less than 7

acid rain – rain with a pH of less than 5 that is produced when pollutants such as sulfer dioxide dissolve in rain water

alkali – a base that dissolves in water; has a pH greater than 7

ammeter – an instrument used to measure electric current

amniotic sac – a bag of liquid that protects the baby from bumps during the mother's everyday activities

ampere (amp) – the unit electric current is measured in

anode – electrode connected to positive terminal of a battery

antacid – a substance that neutralises acids

asthenosphere – layer of the mantle directly below the lithosphere; rocks here flow like thick liquid

atom – the smallest part of an element

B

base – a compound that neutralises an acid; has a pH greater than 7

battery – several electric cells joined together

C

cathode – electrode connected to negative terminal of a battery

cervix – a ring of muscle at the bottom of the uterus that separates the uterus from the vagina

chlamydia – a sexually transmitted infection caused by a type of bacteria

chromosome – a thread-like structure made of DNA that carries the instructions for making living things

circuit (electric) – a pathway along which electrons can move

conception (or fertilisation) – the joining together of male and female sex cells

conductor – a substance that lets an electric current move through it

convection current – movement of hotter, less dense material upwards; cooler, denser material moves downwards

conventional current – current moving in a circuit from the positive to the negative terminal of a battery

convergent – coming together

core – the centre of the earth

corrosive – capable of eating away at something

CPR – cardio-pulmonary resuscitation (mouth-to-mouth)

crust – the solid outer layer of the earth

current – a flow of electrons

D

divergent – moving apart

DNA – deoxyribonucleic acid; the chemical that chromosomes are made of

E

earthquake – shaking of ground caused by sudden movement of rock along a fault

egg – the female gamete or sex cell

ejaculation – the release of sperm out of the penis

electron – negatively charged particle that moves around the nucleus of an atom

embryo – the name given to a baby in the early stage of development less than three months

epicentre – place on the earth's surface directly above the point where an earthquake starts

erosion – the carrying away of broken down rock material

F

fallopian tube (or oviduct) – a tube from the uterus to the womb; the site of fertilisation

fault – a break in the earth's crust

fertilisation (or conception) – the joining together of male and female sex cells

focus – the site inside the earth where an earthquake starts

foetus – the name given to a baby in the later stage of development over three months

G

gamete (sex cell) – sperm and egg

gene – part of a chromosome that carries information to make a protein

genitals – the external sex organs

Gondwanaland – a supercontinent that existed in the southern hemisphere about 200 million years ago

H

hormone – a chemical messenger that travels around the body in the blood

I

indicator – chemical dye that changes colour when acids and bases are added to it

inheritance – the receiving of genetic information from one's parents

insulator – a substance that does not let an electric current move through it

L

lattice – the way in which particles are arranged in a solid

Laurasia – a supercontinent that existed in the northern hemisphere about 200 million years ago

lava – molten magma that has flowed out onto the earth's surface

lithosphere – the crust and upper mantle of the earth

loess – fine mineral particles carried by the wind

M

magma – molten rock inside the earth

mantle – the layer of the earth between the crust and the core

menopause – the time in a woman's life when she stops menstruating

menstruation (period) – the loss of blood and cells from the uterus

Mercalli scale – a scale that measures the effect an earthquake has on buildings

Moment Magnitude scale – a scale that measures the size of an earthquake

moraine – material deposited at the edges of a valley as a glacier moves

N

neutralisation – a reaction between an acid and a base that produces a salt and water

nucleus (of atom) – the heavy, central part of an atom made up of protons and neutrons

O

oestrogen – a hormone produced by the ovaries

ovaries (singular is ovary) – organs in the female body that produce eggs and hormones

oviduct (or fallopian tube) – a tube from the uterus to the womb; the site of fertilisation

P

P-wave – primary seismic wave; the first to be felt in an earthquake

Pangaea – a single supercontinent thought to have existed over 200 million years ago

parallel circuit – circuit a circuit where there is more than one pathway for the current to flow through

penis – male sex organ

pH scale – a scale used to describe how acidic or basic/alkaline a substance is

pituitary gland – a small gland in the brain that produces hormones

placenta – the site of exchange of nutrients, oxygen and wastes between a mother and her developing baby

plaque – a layer of food and bacteria on the surface of the teeth

power (electric) – rate at which electrical energy is transformed into other forms of energy

progesterone – a hormone produced by the corpus luteum in the ovary and by the placenta

prostate gland – a gland that produces a fluid that is mixed with the sperm to form semen

puberty – the age at which a person becomes able to produce children

R

resistance – the opposition of a flow of electrons through a substance

Richter scale – a scale that measures the size of an earthquake

S

S-wave – a secondary seismic wave (slower than P-waves) that can only travel through solids

salinity – saltiness

salt – a substance produced when an acid and a base react together

seismic waves – waves produced by an earthquake

series circuit – a circuit in which there is only one pathway for the current to flow through

sex cells (gametes) – sperm and egg

static electricity – non-moving charges on an object as a result of friction

subduction – the movement of one tectonic plate underneath another plate

T

tectonic plate – pieces of the earth's lithosphere that 'float' on the asthenosphere

testes (singular is testis) – organs in the male body that produce sperm

testosterone – a hormone produced in the testes

U

umbilical cord – a cord connecting a developing baby to the placenta

uterus (womb) – female organ in which a baby develops

V

vagina – a muscular tube between the uterus and the outside of the body

vent – an opening in the surface of the earth through which lava can erupt

viscosity – a measure of how easily a liquid flows

volt – the unit of voltage

voltage – electrical energy

voltmeter – an instrument that measures voltage

vulva – the genitals of a female

W

watt – the unit of power

weathering – the breaking down of rock

womb (uterus) – female organ in which a baby develops

Z

zygote – the cell formed when the sperm and egg join at fertilisation

Printed in Australia
06 Feb 2013
CEN0001F

9 780170 1313